ISEE Upper Level Summer Math Workbook

Essential Summer Learning Math Skills plus Two Complete ISEE Upper Level Math Practice Tests

By

Michael Smith & Reza Nazari

ISEE Upper Level Summer Math Workbook

Published in the United State of America By

The Math Notion

Web: WWW.MathNotion.Com

Email: info@Mathnotion.com

Copyright © 2020 by the Math Notion. All rights reserved. No part of this publication may be reproduced, stored in a retrieval system, or transmitted in any form or by any means, electronic, mechanical, photocopying, recording, scanning, or otherwise, except as permitted under Section 107 or 108 of the 1976 United States Copyright Ac, without permission of the author.

All inquiries should be addressed to the Math Notion.

About the Author

Michael Smith has been a math instructor for over a decade now. He holds a master's degree in Management. Since 2006, Michael has devoted his time to both teaching and developing exceptional math learning materials. As a Math instructor and test prep expert, Michael has worked with thousands of students. He has used the feedback of his students to develop a unique study program that can be used by students to drastically improve their math score fast and effectively.

- **ACT Math Practice Book**
- **SAT Math Practice Book**
- **PSAT Math Practice Book**
- **Accuplacer Math Practice Book**
- **Common Core Math Practice Book**
- **many Math Education Workbooks, Exercise Books and Study Guides**

As an experienced Math teacher, Mr. Smith employs a variety of formats to help students achieve their goals: He tutors online and in person, he teaches students in large groups, and he provides training materials and textbooks through his website and through Amazon.

You can contact Michael via email at:

info@Mathnotion.com

Prepare for The ISEE Upper Level Math Test with a Perfect Workbook!

ISEE Upper Level Summer Math Workbook is a learning math workbook to prevent Summer learning loss. It helps students retain and strengthen their Math skills and provides a strong foundation for success. This workbook provides students with solid foundation to get a head starts on their upcoming school year.

ISEE Upper Level Summer Math Workbook is designed by top test prep experts to help students prepare for the ISEE Upper Level Math test. It provides test-takers with an in-depth focus on the math section of the test, helping them master the essential math skills that test-takers find the most troublesome. This is a prestigious resource for those who need an extra practice to succeed on the ISEE Upper Level Math test in the summer.

ISEE Upper Level Summer Math Workbook contains many exciting and unique features to help your student scores higher on the ISEE Upper Level Math test, including:

- Over 2,500 of standards-aligned math practice questions with answers
- Complete coverage of all Math concepts which students will need to ace the ISEE Upper Level test
- Content 100% aligned with the latest ISEE Upper Level test
- Written by ISEE Upper Level Math experts
- 2 full-length ISEE Upper Level Math practice tests (featuring new question types) with detailed answers

This Comprehensive Summer Workbook for the ISEE Upper Level Math is a perfect resource for those ISEE Upper Level Math test takers who want to review core content areas, brush-up in math, discover their strengths and weaknesses, and achieve their best scores on the ISEE Upper Level test.

WWW.MathNotion.COM

… So Much More Online!

✓ FREE Math Lessons

✓ More Math Learning Books!

✓ Mathematics Worksheets

✓ Online Math Tutors

For a PDF Version of This Book

Please Visit WWW.MathNotion.com

contents

Chapter 1:
Integers and Number Theory

Topics that you will practice in this chapter:

- ✓ Rounding

- ✓ Whole Number Addition and Subtraction

- ✓ Whole Number Multiplication and Division

- ✓ Rounding and Estimates

- ✓ Adding and Subtracting Integers

- ✓ Multiplying and Dividing Integers

- ✓ Order of Operations

- ✓ Ordering Integers and Numbers

- ✓ Integers and Absolute Value

- ✓ Factoring Numbers

- ✓ Greatest Common Factor (GCF)

- ✓ Least Common Multiple (LCM)

"Wherever there is number, there is beauty." –Proclus

Rounding

✎ Round each number to the nearest ten.

1) 52 = ____

2) 89 = ____

3) 34 = ____

4) 66 = ____

5) 42 = ____

6) 78 = ____

7) 121 = ____

8) 91 = ____

9) 48 = ____

10) 21 = ____

11) 134 = ____

12) 157 = ____

✎ Round each number to the nearest hundred.

13) 148 = ____

14) 368 = ____

15) 619 = ____

16) 194 = ____

17) 522 = ____

18) 169 = ____

19) 491 = ____

20) 717 = ____

21) 780 = ____

22) 833 = ____

23) 498 = ____

24) 947 = ____

✎ Round each number to the nearest thousand.

25) 3,325 = ____

26) 2,598 = ____

27) 4,099 = ____

28) 5,808 = ____

29) 6,075 = ____

30) 36,893 = ____

31) 52,199 = ____

32) 80,958 = ____

33) 75,952 = ____

34) 95,250 = ____

35) 78,680 = ____

36) 97,869 = ____

Whole Number Addition and Subtraction

✎ **Find the sum or subtract.**

1)
$$\begin{array}{r} 1{,}982 \\ +\ 895 \\ \hline \end{array}$$

5)
$$\begin{array}{r} 1{,}125 \\ +\ 859.35 \\ \hline \end{array}$$

9)
$$\begin{array}{r} 7{,}322 \\ -\ 895.9 \\ \hline \end{array}$$

2)
$$\begin{array}{r} 3{,}658 \\ -\ 1{,}254 \\ \hline \end{array}$$

6)
$$\begin{array}{r} 857.26 \\ +989.15 \\ \hline \end{array}$$

10)
$$\begin{array}{r} 8{,}921.45 \\ -\ 5{,}214.25 \\ \hline \end{array}$$

3)
$$\begin{array}{r} 582.54 \\ -\ 321.45 \\ \hline \end{array}$$

7)
$$\begin{array}{r} 254.35 \\ +123.89 \\ \hline \end{array}$$

11)
$$\begin{array}{r} 2{,}321.25 \\ +\ 1{,}984.99 \\ \hline \end{array}$$

4)
$$\begin{array}{r} 1{,}254 \\ +\ 852.98 \\ \hline \end{array}$$

8)
$$\begin{array}{r} 3{,}257.5 \\ +1{,}245.2 \\ \hline \end{array}$$

12)
$$\begin{array}{r} 9{,}914.09 \\ -6{,}621.12 \\ \hline \end{array}$$

✎ **Find the missing number.**

13) $362.5 + \underline{\quad} = 985.3$

14) $3{,}856 - \underline{\quad} = 2{,}009.5$

15) $\underline{\quad} - 985.1 = 1{,}450.9$

16) $2{,}785 - 1{,}234.12 = \underline{\quad}$

17) $999.9 + \underline{\quad} = 1{,}234.6$

18) $5{,}758.8 - 3{,}758.85 = \underline{\quad}$

Whole Number Multiplication and Division

✍ **Calculate each product.**

1)
$$\begin{array}{r} 35 \\ \times\,43 \\ \hline \end{array}$$

3)
$$\begin{array}{r} 37.2 \\ \times\,16 \\ \hline \end{array}$$

5)
$$\begin{array}{r} 158.8 \\ \times\,15.4 \\ \hline \end{array}$$

2)
$$\begin{array}{r} 53.2 \\ \times\,12.5 \\ \hline \end{array}$$

4)
$$\begin{array}{r} 27.5 \\ \times\,26 \\ \hline \end{array}$$

6)
$$\begin{array}{r} 143.2 \\ \times\,15.5 \\ \hline \end{array}$$

✍ **Find the missing quotient.**

7) $600 \div 1.5 =$ _____

8) $780 \div 39 =$ _____

9) $390 \div 1.3 =$ _____

10) $900 \div 0.9 =$ _____

11) $156 \div 40 =$ _____

12) $112 \div 1.6 =$ _____

13) $660 \div 2.2 =$ _____

14) $400 \div 0.8 =$ _____

15) $2,040 \div 25.5 =$ _____

16) $9,360 \div 31.2 =$ _____

✍ **Calculate each problem.**

17) $560 \div 7 = N$, $N =$ __

18) $315 \div 4.5 = N$, $N =$ __

19) $N \div 9 = 65$, $N =$ __

20) $24.6 \times N = 147.6$, $N =$ __

21) $985 \div N = 1,970$, $N =$ __

22) $N \times 3.5 = 147$, $N =$ __

Rounding and Estimates

✍ **Estimate the sum by rounding each number to the nearest ten.**

1) $19 + 23 =$ _____

2) $72 + 31 =$ _____

3) $48 + 63 =$ _____

4) $44 + 86 =$ _____

5) $169 + 212 =$ _____

6) $650 + 323 =$ _____

7) $598 + 575 =$ _____

8) $1,586 + 3,355 =$ _____

✍ **Estimate the product by rounding each number to the nearest ten.**

9) $37 \times 43 =$ _____

10) $12 \times 31 =$ _____

11) $48 \times 54 =$ _____

12) $17 \times 33 =$ _____

13) $68 \times 27 =$ _____

14) $91 \times 21 =$ _____

15) $86 \times 37 =$ _____

16) $96 \times 42 =$ _____

✍ **Estimate the sum or product by rounding each number to the nearest ten.**

17)
$$\begin{array}{r} 28 \\ \times\ 16 \\ \hline \underline{\qquad} \end{array}$$

18)
$$\begin{array}{r} 72 \\ \times\ 22 \\ \hline \underline{\qquad} \end{array}$$

19)
$$\begin{array}{r} 85 \\ +\ 64 \\ \hline \underline{\qquad} \end{array}$$

20)
$$\begin{array}{r} 43 \\ +91 \\ \hline \underline{\qquad} \end{array}$$

21)
$$\begin{array}{r} 64 \\ \times\ 39 \\ \hline \underline{\qquad} \end{array}$$

22)
$$\begin{array}{r} 99 \\ +\ 54 \\ \hline \underline{\qquad} \end{array}$$

Adding and Subtracting Integers

✎ **Find each sum.**

1) $15 + (-35) =$

2) $(-28) + (-29) =$

3) $19 + (-27) =$

4) $57 + (-64) =$

5) $(-14) + (-19) + 64 =$

6) $54 + (-36) + 19 =$

7) $46 + (-30) + (-33) + 29 =$

8) $(-40) + (-70) + 28 + 55 =$

9) $60 + (-65) + (83 - 72) =$

10) $49 + (-55) + (90 - 67) =$

✎ **Find each difference.**

11) $(-32) - (-7) =$

12) $40 - (-12) =$

13) $(-60) - 56 =$

14) $27 - (-17) =$

15) $58 - (76 - 29) =$

16) $19 - (-14) - (-22) =$

17) $(39 + 15) - (-46) =$

18) $49 - 17 - (-13) =$

19) $85 - 45 \quad (-18) =$

20) $78 - (-35) - (-63) =$

21) $89 - (-11) - (-26) =$

22) $(19 - 50) - (-95) =$

23) $46 - 49 - (-87) =$

24) $120 - (98 + 24) - (-38) =$

25) $112 - (-102) + (-81) =$

26) $108 - (-42) + (-89) =$

Multiplying and Dividing Integers

✍ **Find each product.**

1) $(-7) \times (-9) =$

2) $(-5) \times 6 =$

3) $10 \times (-15) =$

4) $(-9) \times (-25) =$

5) $(-7) \times (-12) \times 13 =$

6) $(15 - 4) \times (-11) =$

7) $25 \times (-4) \times (-5) =$

8) $(85 + 10) \times (-11) =$

9) $12 \times (-19 + 12) \times 5 =$

10) $(-15) \times (-18) \times (-20) =$

✍ **Find each quotient.**

11) $85 \div (-5) =$

12) $(-90) \div (-15) =$

13) $(-121) \div (-11) =$

14) $99 \div (-33) =$

15) $(-114) \div 2 =$

16) $(-208) \div (-16) =$

17) $198 \div (-11) =$

18) $(-364) \div (-14) =$

19) $255 \div (-15) =$

20) $(-378) \div (18) =$

21) $(-184) \div (-8) =$

22) $-437 \div (-23) =$

23) $(-570) \div (-19) =$

24) $480 \div (-32) =$

25) $(-546) \div (-21) =$

26) $(486) \div (-54) =$

Order of Operations

✎ **Evaluate each expression.**

1) $7 + (5 \times 8) =$

2) $16 - (6 \times 9) =$

3) $(17 \times 5) + 12 =$

4) $(24 - 12) - (11 \times 4) =$

5) $35 + (18 \div 3) =$

6) $(27 \times 3) \div 3 =$

7) $(88 \div 4) \times (-5) =$

8) $(9 \times 9) + (86 - 52) =$

9) $78 + (5 \times 12) + 14 =$

10) $(60 \times 4) \div (4 + 2) =$

11) $(-15) + (14 \times 4) + 18 =$

12) $(14 \times 5) - (56 \div 7) =$

13) $(7 \times 9 \div 3) - (32 + 21) =$

14) $(45 + 11 - 14) \times 2 - 15 =$

15) $(40 - 18 + 20) \times (75 \div 3) =$

16) $75 + \big(54 - (45 \div 9)\big) =$

17) $(12 + 15 - 24) + (44 \div 4) =$

18) $(78 - 19) + (27 - 10 + 7) =$

19) $(18 \times 3) + (17 \times 9) - 52 =$

20) $65 + 17 - (45 \times 2) + 40 =$

Ordering Integers and Numbers

✍ **Order each set of integers from least to greatest.**

1) $17, -15, -8, 0, 9$ ＿＿, ＿＿, ＿＿, ＿＿, ＿＿, ＿＿

2) $-14, -26, 17, 42, 39$ ＿＿, ＿＿, ＿＿, ＿＿, ＿＿, ＿＿

3) $32, -15, -69, 41, -80$ ＿＿, ＿＿, ＿＿, ＿＿, ＿＿, ＿＿

4) $-49, -65, 35, -21, 68$ ＿＿, ＿＿, ＿＿, ＿＿, ＿＿, ＿＿

5) $69, -32, 10, -45, 24$ ＿＿, ＿＿, ＿＿, ＿＿, ＿＿, ＿＿

6) $108, 76, -59, 87, -78$ ＿＿, ＿＿, ＿＿, ＿＿, ＿＿, ＿＿

✍ **Order each set of integers from greatest to least.**

7) $62, 98, -7, -19, -1$ ＿＿, ＿＿, ＿＿, ＿＿, ＿＿, ＿＿

8) $34, 35, -24, -46, 56$ ＿＿, ＿＿, ＿＿, ＿＿, ＿＿, ＿＿

9) $35, -96, -58, 17, -34$ ＿＿, ＿＿, ＿＿, ＿＿, ＿＿, ＿＿

10) $37, 12, -26, -13, 52$ ＿＿, ＿＿, ＿＿, ＿＿, ＿＿, ＿＿

11) $-12, 66, -18, -28, 54$ ＿＿, ＿＿, ＿＿, ＿＿, ＿＿, ＿＿

12) $-100, -85, -30, 5, 9$ ＿＿, ＿＿, ＿＿, ＿＿, ＿＿, ＿＿

Integers and Absolute Value

✎ **Write absolute value of each number.**

1) $|-19| =$

2) $|-32| =$

3) $|-50| =$

4) $|31| =$

5) $|57| =$

6) $|-76| =$

7) $|42| =$

8) $|101| =$

9) $|28| =$

10) $|-49| =$

11) $|-13|$

12) $|78| =$

13) $|100| =$

14) $|0| =$

15) $|-105| =$

16) $|-77| =$

17) $88 =$

18) $|-29| =$

19) $|112| =$

20) $|-120| =$

✎ **Evaluate the value.**

21) $|-5| - \dfrac{|-40|}{8} =$

22) $18 - |4 - 19| - |-15| =$

23) $\dfrac{|-72|}{9} \times |-9| =$

24) $\dfrac{|6 \times (-8)|}{3} \times \dfrac{|-21|}{7} =$

25) $|5 \times (-9)| + \dfrac{|-110|}{11} =$

26) $\dfrac{|-96|}{12} \times \dfrac{|-27|}{9} =$

27) $|-19 + 12| \times \dfrac{|-12 \times 13|}{7} =$

28) $\dfrac{|-19 \times 6|}{3} \times |-11| =$

Factoring Numbers

✎ **List all positive factors of each number.**

1) 6	9) 52	17) 90
2) 21	10) 63	18) 93
3) 28	11) 70	19) 95
4) 26	12) 72	20) 96
5) 46	13) 78	21) 98
6) 45	14) 80	22) 102
7) 48	15) 82	23) 124
8) 50	16) 88	24) 125

Greatest Common Factor

✎ **Find the GCF for each number pair.**

1) 6, 2	9) 9, 15	17) 28, 21
2) 8, 4	10) 4, 18	18) 56, 72
3) 5, 3	11) 14, 18	19) 34, 51
4) 6, 4	12) 25, 30	20) 6, 18, 27
5) 7, 5	13) 27, 45	21) 2, 9, 8
6) 8, 18	14) 36, 18	22) 10, 12, 24
7) 14, 21	15) 9, 12	23) 5, 14, 21
8) 6, 14	16) 11, 8	24) 72, 9, 18

Least Common Multiple

✍ **Find the LCM for each number pair.**

1) 6, 5

2) 8, 18

3) 9, 15

4) 15, 20

5) 20, 25

6) 22, 33

7) 6, 28

8) 8, 14

9) 21, 28

10) 14, 28

11) 9, 30

12) 7, 12

13) 12, 36

14) 9, 54

15) 42, 21

16) 40, 16

17) 12, 42

18) 13, 11

19) 32, 72

20) 15, 27

21) 24, 44

22) 8, 12, 42

23) 2, 6, 11

24) 15, 25, 30

Answers of Worksheets – Chapter 1

Rounding

1) 50	10) 20	19) 500	28) 6,000
2) 90	11) 130	20) 700	29) 6,000
3) 30	12) 160	21) 800	30) 37,000
4) 70	13) 100	22) 800	31) 52,000
5) 40	14) 400	23) 500	32) 81,000
6) 80	15) 600	24) 900	33) 76,000
7) 120	16) 200	25) 3,000	34) 95,000
8) 90	17) 500	26) 3,000	35) 79,000
9) 50	18) 200	27) 4,000	36) 98,000

Whole Number Addition and Subtraction

1) 2,877	7) 378.24	13) 622.8
2) 2,404	8) 4,502.7	14) 1,846.5
3) 261.09	9) 6,426.1	15) 2,436
4) 2,106.98	10) 3,707.2	16) 1,550.88
5) 1,984.35	11) 4,306.24	17) 234.7
6) 1,846.41	12) 3,292.97	18) 1,999.95

Whole Number Multiplication and Division

1) 1,505	7) 400	13) 300	19) 585
2) 665	8) 20	14) 500	20) 6
3) 595.2	9) 300	15) 80	21) 2
4) 715	10) 1,000	16) 300	22) 42
5) 2,445.52	11) 3.9	17) 80	
6) 2,219.6	12) 70	18) 70	

Rounding and Estimates

1) 40	6) 970	11) 2,500	16) 4,200
2) 100	7) 1,180	12) 600	17) 600
3) 110	8) 4,950	13) 2,100	18) 1,400
4) 130	9) 1,600	14) 1,800	19) 150
5) 380	10) 300	15) 3,600	20) 130

21) 2,400   22) 150

Adding and Subtracting Integers

1) −20	8) −27	15) 11	22) 64
2) −57	9) 6	16) 55	23) 84
3) −8	10) 17	17) 100	24) 36
4) −7	11) −25	18) 45	25) 133
5) 31	12) 52	19) 58	26) 61
6) 37	13) −116	20) 176	
7) 12	14) 44	21) 126	

Multiplying and Dividing Integers

1) 63	8) −1,045	15) −57	22) 19
2) −30	9) −420	16) 13	23) 30
3) −150	10) −5,400	17) −18	24) −15
4) 225	11) −17	18) 26	25) 26
5) 1,092	12) 6	19) −17	26) −9
6) −121	13) 11	20) −21	
7) 500	14) −3	21) 23	

Order of Operations

1) 47	6) 27	11) 59	16) 124
2) −38	7) −110	12) 62	17) 14
3) 97	8) 115	13) −32	18) 83
4) −32	9) 152	14) 69	19) 155
5) 41	10) 40	15) 1,050	20) 32

Ordering Integers and Numbers

1) −15, −8, 0, 9, 17

2) −26, −14, 17, 39, 42

3) −80, −69, −15, 32, 41

4) −65, −49, −21, 35, 68

5) −45, −32, 10, 24, 69

6) −78, −59, 76, 87, 108

7) 98, 62, −1, −7, −19

8) 56, 35, 34, −24, −46

9) 35, 17, −34, −58, −96

10) 52, 37, 12, −13, −26

11) 66, 54, −12, −18, −28

12) 9, 5, −30, −85, −100

Integers and Absolute Value

1) 19	8) 101	15) 105	22) -12
2) 32	9) 28	16) 77	23) 72
3) 50	10) 49	17) 88	24) 48
4) 31	11) 13	18) 29	25) 55
5) 57	12) 78	19) 112	26) 24
6) 76	13) 100	20) 120	27) 156
7) 42	14) 0	21) 0	28) 418

Factoring Numbers

1) 1, 2, 3, 6

2) 1, 3, 7, 21

3) 1, 2, 4, 7, 14, 28

4) 1, 2, 13, 26

5) 1, 2, 23, 46

6) 1, 3, 5, 9, 15, 45

7) 1, 2, 3, 4, 6, 8, 12, 16, 24, 48

8) 1, 2, 5, 10, 25, 50

9) 1, 2, 4, 5, 13, 26, 52

10) 1, 3, 7, 9, 21, 63

11) 1, 2, 5, 7, 10, 14, 35, 70

12) 1, 2, 3, 4, 6, 8, 9, 12, 18 24, 36, 72

13) 1, 2, 3, 6, 13, 26, 39, 78

14) 1, 2, 4, 5, 8, 10, 16, 20, 40, 80

15) 1, 2, 41, 82

16) 1, 2, 4, 8, 11, 22, 44, 88

17) 1, 2, 3, 5, 6, 9, 10, 15, 18, 30, 45, 90

18) 1, 3, 31, 93

19) 1, 5, 19, 95

20) 1, 2, 3, 4, 6, 8, 12, 16, 24, 32, 48, 96

21) 1, 2, 7, 14, 49, 98

22) 1, 2, 3, 6, 17, 34, 51, 102

23) 1, 2, 4, 31, 62, 124

24) 1, 5, 25, 125

Greatest Common Factor

1) 2	7) 7	13) 9	19) 17
2) 4	8) 2	14) 18	20) 3
3) 1	9) 3	15) 3	21) 1
4) 2	10) 2	16) 1	22) 2
5) 1	11) 2	17) 7	23) 1
6) 2	12) 5	18) 8	24) 9

Least Common Multiple

1) 30	7) 84	13) 36	19) 288
2) 72	8) 56	14) 54	20) 135
3) 45	9) 84	15) 42	21) 264
4) 60	10) 28	16) 80	22) 168
5) 100	11) 90	17) 84	23) 66
6) 66	12) 84	18) 143	24) 150

Chapter 2:

Fractions and Decimals

Topics that you will practice in this chapter:

- ✓ Simplifying Fractions
- ✓ Adding and Subtracting Fractions
- ✓ Multiplying and Dividing Fractions
- ✓ Adding and Subtract Mixed Numbers
- ✓ Multiplying and

- Dividing Mixed Numbers
- ✓ Adding and Subtracting Decimals
- ✓ Multiplying and Dividing Decimals
- ✓ Comparing Decimals
- ✓ Rounding Decimals

"A Man is like a fraction whose numerator is what he is and whose denominator is what he thinks of himself. The larger the denominator, the smaller the fraction." –Tolstoy

Simplifying Fractions

✎ **Simplify each fraction to its lowest terms.**

1) $\frac{8}{16} =$

2) $\frac{28}{35} =$

3) $\frac{27}{36} =$

4) $\frac{70}{140} =$

5) $\frac{13}{52} =$

6) $\frac{38}{57} =$

7) $\frac{64}{80} =$

8) $\frac{21}{84} =$

9) $\frac{85}{170} =$

10) $\frac{120}{168} =$

11) $\frac{31}{124} =$

12) $\frac{48}{96} =$

13) $\frac{98}{112} =$

14) $\frac{99}{110} =$

15) $\frac{51}{153} =$

16) $\frac{40}{112} =$

17) $\frac{90}{225} =$

18) $\frac{44}{297} =$

19) $\frac{54}{279} =$

20) $\frac{320}{720} =$

21) $\frac{70}{560} =$

✎ **Find the answer for each problem.**

22) Which of the following fractions equal to $\frac{3}{7}$? _____

    A. $\frac{24}{63}$ B. $\frac{51}{109}$ C. $\frac{51}{119}$ D. $\frac{240}{630}$

23) Which of the following fractions equal to $\frac{7}{8}$? _____

    A. $\frac{182}{208}$ B. $\frac{175}{208}$ C. $\frac{182}{216}$ D. $\frac{49}{64}$

24) Which of the following fractions equal to $\frac{2}{9}$? _____

    A. $\frac{64}{126}$ B. $\frac{46}{207}$ C. $\frac{48}{207}$ D. $\frac{56}{208}$

Adding and Subtracting Fractions

✎ Find the sum.

1) $\dfrac{5x}{8} + \dfrac{3x}{8} =$

2) $\dfrac{x}{2} + \dfrac{x}{7} =$

3) $\dfrac{y}{3} + \dfrac{y}{4} =$

4) $\dfrac{3x}{8} + \dfrac{2x}{5} =$

5) $\dfrac{xy}{5} + \dfrac{2xy}{7} =$

6) $\dfrac{2x}{9} + \dfrac{4x}{9} =$

7) $\dfrac{a}{4} + \dfrac{2a}{3} =$

8) $\dfrac{2}{x} + \dfrac{4}{x} =$

9) $\dfrac{1}{a} + \dfrac{2}{b} =$

10) $\dfrac{3b}{5} + \dfrac{2b}{7} =$

11) $\dfrac{a}{y} + \dfrac{3a}{y} =$

12) $\dfrac{3}{x} + \dfrac{1}{2x} =$

✎ Find the difference.

13) $\dfrac{x}{3} - \dfrac{x}{6} =$

14) $\dfrac{2x}{5} - \dfrac{3x}{8} =$

15) $\dfrac{x}{7} - \dfrac{y}{7} =$

16) $\dfrac{2x}{7} - \dfrac{x}{6} =$

17) $\dfrac{5a}{9} - \dfrac{2a}{5} =$

18) $\dfrac{2ab}{3} - \dfrac{ab}{6} =$

19) $\dfrac{1}{x} - \dfrac{1}{3x} =$

20) $\dfrac{4}{y} - \dfrac{3}{4y} =$

21) $\dfrac{5}{x} - \dfrac{2y}{xy} =$

22) $\dfrac{8}{ab} - \dfrac{5}{3ab} =$

23) $\dfrac{2a}{y} - \dfrac{a}{3y} =$

24) $\dfrac{5}{b} - \dfrac{2}{3b} =$

25) $\dfrac{2a}{b} - \dfrac{a}{b} =$

26) $\dfrac{3}{a} - \dfrac{2}{b} =$

27) $\dfrac{4a}{b} - \dfrac{2a}{3b} =$

28) $\dfrac{6}{xy} - \dfrac{7}{2xy} =$

29) $\dfrac{2}{a} - \dfrac{1}{4a} =$

30) $\dfrac{2a}{3b} - \dfrac{4a}{9b} =$

Multiplying and Dividing Fractions

✎ **Find the value of each expression in lowest terms.**

1) $\dfrac{3}{a} \times \dfrac{5}{3} =$

2) $\dfrac{2}{3b} \times \dfrac{9}{2} =$

3) $\dfrac{a}{15} \times \dfrac{5}{2a} =$

4) $\dfrac{x}{3a} \times \dfrac{9a}{6x} =$

5) $\dfrac{x}{12} \times \dfrac{6}{y} =$

6) $\dfrac{7}{x} \times \dfrac{x}{14} =$

7) $\dfrac{10}{3a} \times \dfrac{6}{20} =$

8) $\dfrac{4a}{b} \times \dfrac{2b}{5} =$

9) $\dfrac{2ab}{7} \times \dfrac{14}{6ab} =$

10) $\dfrac{4a}{5b} \times \dfrac{15}{2a} =$

11) $\dfrac{ab}{21} \times \dfrac{7}{a} =$

12) $\dfrac{a}{cd} \times \dfrac{2bc}{a} =$

✎ **Find the value of each expression in lowest terms.**

13) $\dfrac{a}{2} \div \dfrac{a}{4} =$

14) $\dfrac{b}{3} \div \dfrac{b}{9} =$

15) $\dfrac{a}{b} \div \dfrac{3}{b} =$

16) $\dfrac{2a}{15} \div \dfrac{4a}{5} =$

17) $\dfrac{1}{a} \div \dfrac{b}{3a} =$

18) $\dfrac{4a}{3b} \div \dfrac{a}{2b} =$

19) $\dfrac{a}{8} \div \dfrac{3a}{16} =$

20) $\dfrac{3b}{20} \div \dfrac{6b}{15a} =$

21) $\dfrac{x}{12y} \div \dfrac{2x}{9y} =$

22) $\dfrac{25}{x} \div \dfrac{50}{2x} =$

23) $\dfrac{16}{5ab} \div \dfrac{32}{ab} =$

24) $\dfrac{7a}{b} \div \dfrac{8a}{b} =$

25) $\dfrac{5}{x} \div \dfrac{3y}{x} =$

26) $\dfrac{2a}{21} \div \dfrac{a}{14} =$

27) $\dfrac{ab}{x} \div \dfrac{a}{x} =$

28) $\dfrac{6}{a} \div \dfrac{3b}{2a} =$

29) $\dfrac{9}{16a} \div \dfrac{3}{8ab} =$

30) $\dfrac{24}{xy} \div \dfrac{12}{y} =$

Adding and Subtracting Mixed Numbers

✍ **Find the sum.**

1) $3\frac{5}{6} + 2\frac{1}{3} =$

2) $4\frac{2}{5} + 1\frac{1}{5} =$

3) $5\frac{1}{8} + 6\frac{3}{4} =$

4) $2\frac{2}{3} + 3\frac{1}{2} =$

5) $3\frac{4}{5} + 3\frac{2}{15} =$

6) $8\frac{1}{16} + 3\frac{3}{8} =$

7) $4\frac{3}{5} + 4\frac{1}{6} =$

8) $7\frac{3}{4} + 3\frac{5}{6} =$

9) $8\frac{5}{6} + 2\frac{2}{7} =$

10) $11\frac{3}{16} + 3\frac{5}{24} =$

✍ **Find the difference.**

11) $3\frac{3}{4} - 2\frac{1}{4} =$

12) $5\frac{1}{7} - 3\frac{1}{7} =$

13) $4\frac{1}{3} - 1\frac{1}{9} =$

14) $7\frac{1}{6} - 3\frac{1}{12} =$

15) $6\frac{1}{3} - 2\frac{5}{18} =$

16) $8\frac{1}{4} - 5\frac{1}{8} =$

17) $9\frac{1}{2} - 6\frac{1}{5} =$

18) $11\frac{7}{15} - 8\frac{1}{30} =$

19) $12\frac{3}{5} - 7\frac{2}{7} =$

20) $18\frac{1}{8} - 14\frac{3}{16} =$

21) $12\frac{2}{3} - 11\frac{7}{15} =$

22) $3\frac{1}{5} - 1\frac{1}{2} =$

23) $14\frac{3}{5} - 6\frac{4}{5} =$

24) $17\frac{1}{4} - 14\frac{8}{9} =$

25) $24\frac{3}{9} - 15\frac{1}{18} =$

26) $28\frac{3}{7} - 19\frac{5}{6} =$

Multiplying and Dividing Mixed Numbers

✎ **Find the product.**

1) $2\frac{1}{3} \times 4\frac{1}{2} =$

2) $4\frac{1}{5} \times 2\frac{1}{3} =$

3) $7\frac{2}{3} \times 3\frac{3}{5} =$

4) $9\frac{2}{7} \times 3\frac{1}{8} =$

5) $5\frac{4}{11} \times 4\frac{1}{3} =$

6) $7\frac{3}{8} \times 5\frac{4}{9} =$

7) $9\frac{2}{3} \times 11\frac{5}{6} =$

8) $8\frac{3}{5} \times 7\frac{4}{9} =$

9) $5\frac{1}{9} \times 9\frac{5}{8} =$

10) $10\frac{2}{7} \times 2\frac{5}{8} =$

✎ **Find the quotient.**

11) $2\frac{1}{8} \div 1\frac{3}{8} =$

12) $4\frac{1}{6} \div 2\frac{1}{3} =$

13) $7\frac{1}{3} \div 3\frac{3}{4} =$

14) $4\frac{5}{8} \div 1\frac{1}{2} =$

15) $6\frac{5}{12} \div 4\frac{1}{6} =$

16) $5\frac{7}{18} \div 5\frac{1}{6} =$

17) $6\frac{5}{21} \div 2\frac{3}{7} =$

18) $8\frac{1}{7} \div 8\frac{1}{14} =$

19) $10\frac{1}{4} \div 3\frac{2}{5} =$

20) $15\frac{1}{3} \div 5\frac{2}{9} =$

21) $12\frac{1}{3} \div 6\frac{1}{2} =$

22) $18\frac{1}{9} \div 18\frac{1}{6} =$

23) $10\frac{3}{4} \div 5\frac{2}{5} =$

24) $11\frac{1}{3} \div 8\frac{4}{5} =$

25) $9\frac{1}{6} \div 3\frac{2}{7} =$

26) $7\frac{1}{3} \div 3\frac{7}{11} =$

Adding and Subtracting Decimals

✎ **Add and subtract decimals.**

1)
$$52.18 - 21.27$$

4)
$$65.84 - 35.49$$

7)
$$98.12 - 45.55$$

2)
$$49.34 + 25.24$$

5)
$$54.57 + 18.37$$

8)
$$48.99 + 57.67$$

3)
$$48.60 + 35.75$$

6)
$$90.45 - 28.75$$

9)
$$158.05 - 78.98$$

✎ **Find the missing number.**

10) ___ $+ 4.9 = 6.5$

11) $5.15 +$ ___ $= 6.43$

12) $8.09 +$ ___ $= 11.84$

13) $8.88 -$ ___ $= 6.78$

14) ___ $- 1.59 = 3.71$

15) ___ $- 19.98 = 8.17$

16) $38.89 +$ ___ $= 41.32$

17) ___ $- 35.99 = 1.80$

18) ___ $+ 39.08 = 41.36$

19) $98.98 +$ ___ $= 123.68$

Multiplying and Dividing Decimals

✎ **Find the product.**

1) $0.6 \times 0.8 =$

2) $2.5 \times 0.9 =$

3) $0.87 \times 0.4 =$

4) $0.15 \times 0.75 =$

5) $0.95 \times 0.7 =$

6) $1.57 \times 0.9 =$

7) $5.85 \times 1.3 =$

8) $12.5 \times 4.5 =$

9) $19.8 \times 7.32 =$

10) $85.1 \times 1.5 =$

11) $79.5 \times 11.2 =$

12) $86.9 \times 21.5 =$

✎ **Find the quotient.**

13) $3.25 \div 10 =$

14) $24.5 \div 100 =$

15) $3.9 \div 3 =$

16) $91.2 \div 0.6 =$

17) $29.2 \div 0.4 =$

18) $38.7 \div `9 =$

19) $297.8 \div 1,000 =$

20) $53.55 \div 0.7 =$

21) $345.45 \div 0.1 =$

22) $70.27 \div 0.25 =$

23) $28.968 \div 0.3 =$

24) $86.34 \div 0.06 =$

Comparing Decimals

✍ **Write the correct comparison symbol (>, < or =).**

1) 0.80 ☐ 0.080

2) 0.086 ☐ 0.86

3) 7.090 ☐ 7.09

4) 3.25 ☐ 3.06

5) 4.09 ☐ 0.490

6) 6.06 ☐ 6.6

7) 6.08 ☐ 6.080

8) 4.05 ☐ 4.2

9) 12.35 ☐ 12.198

10) 0.957 ☐ 0.0957

11) 25.24 ☐ 25.240

12) 0.742 ☐ 0.752

13) 14.09 ☐ 14.10

14) 17.45 ☐ 17.154

15) 11.44 ☐ 11.439

16) 15.41 ☐ 15.410

17) 21.43 ☐ 21.043

18) 8.098 ☐ 8.90

19) 16.044 ☐ 16.040

20) 32.35 ☐ 32.350

Rounding Decimals

✎ **Round each decimal to the nearest whole number.**

1) 56.27   3) 18.32   5) 7.90

2) 5.9   4) 4.8   6) 57.7

✎ **Round each decimal to the nearest tenth.**

7) 42.785   9) 96.586   11) 27.198

8) 15.224   10) 101.78   12) 96.87

✎ **Round each decimal to the nearest hundredth.**

13) 9.648   15) 89.2882   17) 68.229

14) 27.819   16) 120.912   18) 85.642

✎ **Round each decimal to the nearest thousandth.**

19) 19.88486   21) 145.9322   23) 189.0991

20) 46.72611   22) 210.1581   24) 121.76798

Answers of Worksheets – Chapter 2

Simplifying Fractions

1) $\frac{1}{2}$ 7) $\frac{4}{5}$ 13) $\frac{7}{8}$ 19) $\frac{6}{31}$

2) $\frac{4}{5}$ 8) $\frac{1}{4}$ 14) $\frac{9}{10}$ 20) $\frac{4}{9}$

3) $\frac{3}{4}$ 9) $\frac{1}{2}$ 15) $\frac{1}{3}$ 21) $\frac{1}{8}$

4) $\frac{1}{2}$ 10) $\frac{5}{7}$ 16) $\frac{5}{14}$ 22) C

5) $\frac{1}{4}$ 11) $\frac{1}{4}$ 17) $\frac{2}{5}$ 23) A

                                                     24) B

6) $\frac{2}{3}$ 12) $\frac{1}{2}$ 18) $\frac{4}{27}$

Adding and Subtracting Fractions

1) $\frac{8x}{8} = x$ 9) $\frac{a+2b}{ab}$ 17) $\frac{7a}{45}$ 25) $\frac{a}{b}$

2) $\frac{9x}{14}$ 10) $\frac{31b}{35}$ 18) $\frac{ab}{2}$ 26) $\frac{3b-2a}{ab}$

3) $\frac{7x}{12}$ 11) $\frac{4a}{y}$ 19) $\frac{2}{3x}$ 27) $\frac{10a}{3b}$

4) $\frac{31x}{40}$ 12) $\frac{7}{2x}$ 20) $\frac{13}{4y}$ 28) $\frac{5}{2xy}$

5) $\frac{17xy}{35}$ 13) $\frac{x}{6}$ 21) $\frac{3}{x}$ 29) $\frac{7}{4a}$

6) $\frac{2x}{3}$ 14) $\frac{x}{40}$ 22) $\frac{19}{3ab}$ 30) $\frac{2a}{9b}$

7) $\frac{11a}{12}$ 15) $\frac{x-y}{7}$ 23) $\frac{5a}{3y}$

8) $\frac{6}{x}$ 16) $\frac{5x}{42}$ 24) $\frac{13}{3b}$

Multiplying and Dividing Fractions

1) $\frac{5}{a}$ 4) $\frac{1}{2}$ 7) $\frac{1}{a}$

2) $\frac{3}{b}$ 5) $\frac{x}{2y}$ 8) $\frac{8a}{5}$

3) $\frac{1}{6}$ 6) $\frac{1}{2}$ 9) $\frac{2}{3}$

10) $\frac{6}{b}$

11) $\frac{b}{3}$

12) $\frac{2b}{d}$

13) 2

14) 3

15) $\frac{a}{3}$

16) $\frac{1}{6}$

17) $\frac{3}{b}$

18) $\frac{8}{3}$

19) $\frac{2}{3}$

20) $\frac{3a}{8}$

21) $\frac{3}{8}$

22) 1

23) $\frac{1}{10}$

24) $\frac{7}{8}$

25) $\frac{5}{3y}$

26) $\frac{4}{3}$

27) b

28) $\frac{4}{b}$

29) $\frac{3b}{2}$

30) $\frac{2}{x}$

Adding and Subtracting Mixed Numbers

1) $6\frac{1}{6}$

2) $5\frac{3}{5}$

3) $11\frac{7}{8}$

4) $6\frac{1}{6}$

5) $6\frac{14}{15}$

6) $11\frac{7}{16}$

7) $8\frac{23}{30}$

8) $11\frac{7}{12}$

9) $11\frac{5}{42}$

10) $14\frac{19}{48}$

11) $1\frac{1}{2}$

12) 2

13) $3\frac{2}{9}$

14) $4\frac{1}{12}$

15) $4\frac{1}{18}$

16) $3\frac{1}{8}$

17) $3\frac{3}{10}$

18) $3\frac{13}{30}$

19) $5\frac{11}{35}$

20) $3\frac{15}{16}$

21) $1\frac{1}{5}$

22) $1\frac{7}{10}$

23) $7\frac{4}{5}$

24) $2\frac{13}{36}$

25) $9\frac{5}{18}$

26) $8\frac{25}{42}$

Multiplying and Dividing Mixed Numbers

1) $10\frac{1}{2}$

2) $9\frac{4}{5}$

3) $27\frac{3}{5}$

4) $29\frac{1}{56}$

5) $23\frac{8}{33}$

6) $40\frac{11}{72}$

7) $144\frac{7}{18}$

8) $64\frac{1}{45}$

9) $49\frac{7}{36}$

10) 27

11) $1\frac{6}{11}$

12) $1\frac{11}{14}$

13) $1\frac{43}{45}$

14) $3\frac{1}{12}$

15) $1\frac{27}{50}$

16) $1\frac{4}{93}$

17) $2\frac{29}{51}$

18) $1\frac{1}{113}$

19) $3\frac{1}{68}$

20) $2\frac{44}{47}$

21) $1\frac{35}{39}$

22) $\frac{326}{327}$

23) $1\frac{107}{108}$

24) $1\frac{19}{66}$

25) $2\frac{109}{138}$

26) $2\frac{1}{60}$

Adding and Subtracting Decimals

1) 30.91
2) 74.58
3) 84.35
4) 30.35
5) 72.94
6) 61.7
7) 52.57
8) 106.66
9) 79.07
10) 1.6
11) 1.28
12) 3.75
13) 2.1
14) 5.3
15) 28.15
16) 2.43
17) 37.79
18) 2.28
19) 24.7

Multiplying and Dividing Decimals

1) 0.48
2) 2.25
3) 0.348
4) 0.1125
5) 0.665
6) 1.413
7) 7.605
8) 56.25
9) 144.936
10) 127.65
11) 890.4
12) 1,868.35
13) 0.325
14) 0.245
15) 1.3
16) 152
17) 73
18) 4.3
19) 0.2978
20) 76.5
21) 3,454.5
22) 281.08
23) 96.56
24) 1,439

Comparing Decimals

1) >
2) <
3) =
4) >
5) >
6) <
7) =
8) <
9) >
10) >
11) =
12) <
13) <
14) >
15) >
16) =
17) >
18) <
19) >
20) =

Rounding Decimals

1) 56
2) 6
3) 18
4) 5
5) 8
6) 58
7) 42.8
8) 15.2
9) 96.6
10) 101.8
11) 27.2
12) 96.9
13) 9.65
14) 27.82
15) 89.29
16) 120.91
17) 68.23
18) 85.64
19) 19.885
20) 46.726
21) 145.932
22) 210.158
23) 189.099
24) 121.768

Chapter 3:
Proportions, Ratios, and Percent

Topics that you will practice in this chapter:

- ✓ Simplifying Ratios
- ✓ Proportional Ratios
- ✓ Similarity and Ratios
- ✓ Ratio and Rates Word Problems
- ✓ Percentage Calculations
- ✓ Percent Problems
- ✓ Discount, Tax and Tip
- ✓ Percent of Change
- ✓ Simple Interest

Without mathematics, there's nothing you can do. Everything around you is mathematics. Everything around you is numbers." – Shakuntala Devi

Simplifying Ratios

✍ **Reduce each ratio.**

1) $15:20 =$ ___ : ___

2) $9:90 =$ ___ : ___

3) $24:42 =$ ___ : ___

4) $7:21 =$ ___ : ___

5) $11:110$ ___ : ___

6) $8:64 =$ ___ : ___

7) $18:72 =$ ___ : ___

8) $10:25 =$ ___ : ___

9) $7:42 =$ ___ : ___

10) $49:63 =$ ___ : ___

11) $12:18$ ___ : ___

12) $35:10$ ___ : ___

13) $150:15$ ___ : ___

14) $2.4:3.2$ ___ : ___

15) $7:56 =$ ___ : ___

16) $45:63$ ___ : ___

17) $77:99$ ___ : ___

18) $39:13$ ___ : ___

19) $15:45$ ___ : ___

20) $84:12$ ___ : ___

21) $25:5$ ___ : ___

22) $70:56$ ___ : ___

23) $70:140$ ___ : ___

24) $1.2:36$ ___ : ___

✍ **Write each ratio as a fraction in simplest form.**

25) $7:14 =$

26) $27:45 =$

27) $24:56 =$

28) $16:48 =$

29) $22:66 =$

30) $21:98 =$

31) $34:68 =$

32) $6:30 =$

33) $35:84 =$

34) $12:54 =$

35) $88:104 =$

36) $36:81 =$

37) $1.5:18 =$

38) $4.5:16.5 =$

39) $5:75 =$

40) $3.1:12.4 =$

41) $1.6:6.4 =$

42) $0.25:1.25 =$

43) $8.8:16.4 =$

44) $0.75:6.75 =$

45) $1.8:3 =$

Proportional Ratios

✎ **Fill in the blanks; Calculate each proportion.**

1) $3:8 = \underline{\quad} : 32$

2) $1:2 = 45:\underline{\quad}$

3) $1:11 = \underline{\quad} : 55$

4) $9:12 = 18:\underline{\quad}$

5) $9:7 = 81:\underline{\quad}$

6) $2:8 = \underline{\quad} : 56$

7) $2.3:1.2 = \underline{\quad} : 12$

8) $0.5:2 = \underline{\quad} : 32$

9) $1.6:2 = \underline{\quad} : 60$

10) $2.5:4.5 = \underline{\quad} : 90$

11) $3.8:7.1 = 7.6:\underline{\quad}$

12) $5.5:6 = 16.5:\underline{\quad}$

✎ **State if each pair of ratios form a proportion.**

13) $\frac{5}{12}$ and $\frac{15}{36}$

14) $\frac{2}{4}$ and $\frac{18}{36}$

15) $\frac{7}{8}$ and $\frac{28}{32}$

16) $\frac{3}{8}$ and $\frac{27}{64}$

17) $\frac{1}{14}$ and $\frac{5}{65}$

18) $\frac{7}{11}$ and $\frac{70}{100}$

19) $\frac{12}{15}$ and $\frac{48}{60}$

20) $\frac{3}{17}$ and $\frac{36}{204}$

21) $\frac{1.2}{1.5}$ and $\frac{1.44}{22.5}$

22) $\frac{1.3}{1.1}$ and $\frac{3.9}{33}$

23) $\frac{0.7}{0.9}$ and $\frac{6.3}{8.1}$

24) $\frac{2.4}{3.2}$ and $\frac{48}{64}$

✎ **Calculate each proportion.**

25) $\frac{14}{16} = \frac{21}{x}, x = \underline{\quad}$

26) $\frac{3}{28} = \frac{42}{x}, x = \underline{\quad}$

27) $\frac{19}{5} = \frac{38}{x}, x = \underline{\quad}$

28) $\frac{3}{10} = \frac{x}{140}, x = \underline{\quad}$

29) $\frac{4}{9} = \frac{x}{108}, x = \underline{\quad}$

30) $\frac{7}{32} = \frac{21}{x}, x = \underline{\quad}$

31) $\frac{9}{8} = \frac{108}{x}, x = \underline{\quad}$

32) $\frac{12}{17} = \frac{48}{x}, x = \underline{\quad}$

33) $\frac{1.4}{5} = \frac{x}{30}, x = \underline{\quad}$

34) $\frac{1.6}{12} = \frac{x}{60}, x = \underline{\quad}$

35) $\frac{3.5}{15} = \frac{x}{315}, x = \underline{\quad}$

36) $\frac{4.7}{2.5} = \frac{x}{50}, x = \underline{\quad}$

Similarity and Ratios

✎ **Each pair of figures is similar. Find the missing side.**

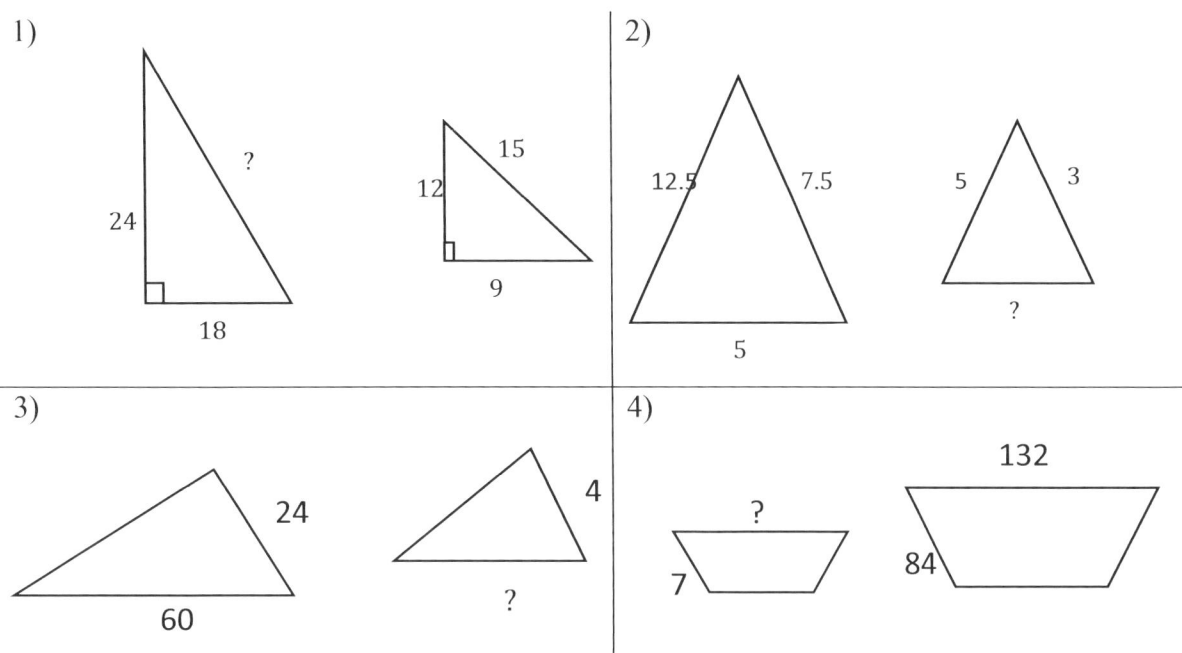

1) ? 24 18 15 12 9

2) 12.5 7.5 5 5 3 ?

3) 24 60 4 ?

4) ? 7 132 84

✎ **Calculate.**

5) Two rectangles are similar. The first is 14 feet wide and 70 feet long. The second is 30 feet wide. What is the length of the second rectangle? _____

6) Two rectangles are similar. One is 3.2 meters by 15 meters. The longer side of the second rectangle is 42 meters. What is the other side of the second rectangle? _____

7) A building casts a shadow 24 ft long. At the same time a girl 10 ft tall casts a shadow 6 ft long. How tall is the building? _____

8) The scale of a map of Texas is 8 inches: 52 miles. If you measure the distance from Dallas to Martin County as 28.8 inches, approximately how far is Martin County from Dallas? _____

Ratio and Rates Word Problems

✎ **Find the answer for each word problem.**

1) Mason has 32 red cards and 40 green cards. What is the ratio of Mason 's red cards to his green cards? _____

2) In a party, 24 soft drinks are required for every 42 guests. If there are 378 guests, how many soft drinks is required? _____

3) In Mason's class, 54 of the students are tall and 30 are short. In Michael's class 126 students are tall and 70 students are short. Which class has a higher ratio of tall to short students? _____

4) The price of 4 apples at the Quick Market is $3.65. The price of 6 of the same apples at Walmart is $4.25. Which place is the better buy? _____

5) The bakers at a Bakery can make 90 bagels in 3 hours. How many bagels can they bake in 17 hours? What is that rate per hour? _____

6) You can buy 8 cans of green beans at a supermarket for $5.60. How much does it cost to buy 56 cans of green beans? _____

7) The ratio of boys to girls in a class is 4: 7. If there are 16 boys in the class, how many girls are in that class? _____

8) The ratio of red marbles to blue marbles in a bag is 3: 4. If there are 42 marbles in the bag, how many of the marbles are red? _____

Percentage Calculations

✍ Calculate the given percent of each value.

1) 3% *of* 60 = ____

2) 20% *of* 80 = ____

3) 25% *of* 80 = ____

4) 24% *of* 50 = ____

5) 18% *of* 150 = ____

6) 70% *of* 35 = ____

7) 15% *of* 28 = ____

8) 32% *of* 300 = ____

9) 54% *of* 80 = ____

10) 10% *of* 610 = ____

11) 35% *of* 520 = ____

12) 64% *of* 110 = ____

13) 44% *of* 200 = ____

14) 28% *of* 94 = ____

15) 30% *of* 85 = ____

16) 68% *of* 102 = ____

17) 45% *of* 160 = ____

18) 55% *of* 220 = ____

✍ Calculate the percent of each given value.

19) ____% *of* 18 = 9

20) ____% *of* 50 = 40

21) ____% *of* 140 = 7

22) ____% *of* 158 = 39.5

23) ____% *of* 75 = 9.375

24) ____% *of* 45 = 11.25

25) ____% *of* 90 = 22.5

26) ____% *of* 650 = 19.5

27) ____% *of* 480 = 24

28) ___% of 400 = 57.32

✍ Calculate each percent problem.

29) A Cinema has 132 seats. 92 seats were sold for the current movie. What percent of seats are empty? _____ %

30) There are 52 boys and 68 girls in a class. 55.00% of the students in the class take the bus to school. How many students do not take the bus to school? _____

Percent Problems

✍ **Calculate each problem.**

1) 30 is what percent of 60? ____%

2) 32 is what percent of 80? ____%

3) 72 is what percent of 45? ____%

4) 8 is what percent of 200? ____%

5) 9 is what percent of 600? ____%

6) 30 is what percent of 500? ____%

7) 70 is what percent of 350? ____%

8) 44 is what percent of 550? ____%

9) 270 is what percent of 900? ____%

10) 180 is what percent of 720? ___%

11) 37.5 is what percent of 75? ___%

12) 27.5 is what percent of 55? ___%

13) 60 is what percent of 750? ___%

14) 22.5 is what percent of 18? ___%

15) 36 is what percent of 24? ___%

16) 18 is what percent of 60? ___%

17) 140 is what percent of 280? ___%

18) 128 is what percent of 40? ___%

✍ **Calculate each percent word problem.**

19) There are 32 employees in a company. On a certain day, 24 were present. What percent showed up for work? _____%

20) A metal bar weighs 36 ounces. 40% of the bar is gold. How many ounces of gold are in the bar? _____

21) A crew is made up of 12 women; the rest are men. If 20% of the crew are women, how many people are in the crew? _____

22) There are 32 students in a class and 8 of them are girls. What percent are boys? _____%

23) The Royals softball team played 310 games and won 248 of them. What percent of the games did they lose? _____%

Discount, Tax and Tip

✍ Find the selling price of each item.

1) Original price of a computer: $450

   Tax: 8% Selling price: $_____

2) Original price of a laptop: $240

   Tax: 4% Selling price: $_____

3) Original price of a sofa: $900

   Tax: 12%   Selling price: $_____

4) Original price of a car: $10,400

   Tax: 2.5%   Selling price: $_____

5) Original price of a Table: $400

   Tax: 3% Selling price: $_____

6) Original price of a house: $360,000

   Tax: 2.8%   Selling price: $_____

7) Original price of a tablet: $150

   Discount: 24%   Selling price: $____

8) Original price of a chair: $180

   Discount: 20%   Selling price: $____

9) Original price of a book: $80

   Discount: 30%   Selling price: $____

10) Original price of a cellphone: $800

    Discount: 20%   Selling price: $___

11) Food bill: $56

    Tip:15%   Price: $_____

12) Food bill: $50

    Tipp: 10%   Price: $_____

13) Food bill: $94

    Tip: 25%   Price: $_____

14) Food bill: $48

    Tipp: 30%   Price: $_____

✍ Find the answer for each word problem.

15) Nicolas hired a moving company. The company charged $400 for its services, and Nicolas gives the movers a 20% tip. How much does Nicolas tip the movers? $_____

16) Mason has lunch at a restaurant and the cost of his meal is $80. Mason wants to leave a 20% tip. What is Mason's total bill including tip? $_____

17) The sales tax in Texas is 14.45% and an item costs $300. How much is the tax? $_____

18) The price of a table at Best Buy is $520. If the sales tax is 4%, what is the final price of the table including tax? $_____

Percent of Change

🖎 Find each percent of change.

1) From 300 to 600.   ___ %

2) From 45 ft to 225 ft.   ___ %

3) From $60 to $420. ___ %

4) From 30 cm to 120 cm. ___ %

5) From 10 to 30. ___ %

6) From 12 to 30.   ___ %

7) From 140 to 210. ___ %

8) From 800 to 400. ___ %

9) From 85 to 51. ___ %

10) From 152 to 76.   ___ %

🖎 Calculate each percent of change word problem.

11) Bob got a raise, and his hourly wage increased from $32 to $40. What is the percent increase? ____ %

12) The price of a pair of shoes increases from $70 to $112. What is the percent increase? ___ %

13) At a coffee shop, the price of a cup of coffee increased from $1.90 to $2.28. What is the percent increase in the cost of the coffee? _____ %

14) 30 cm are cut from a 120 cm board. What is the percent decrease in length? _____ %

15) In a class, the number of students has been increased from 54 to 81. What is the percent increase? _____ %

16) The price of gasoline rose from $22.4 to $25.76 in one month. By what percent did the gas price rise? _____ %

17) A shirt was originally priced at $19. It went on sale for $22.80. What was the percent that the shirt was discounted? _____ %

Simple Interest

✍ **Determine the simple interest for these loans.**

1) $210 at 15% for 4 years. $ _____

6) $28,000 at 3.5% for 6 years. $ _____

2) $1,200 at 6% for 3 years. $ _____

7) $9,600 at 8% for 2 years. $ _____

3) $950 at 25% for 2 years. $ _____

8) $500 at 4.2% for 5 years. $ _____

4) $6,500 at 1.5% for 7 months. $ ____

9) $700 at 2.8 % for 6 months. $ _____

5) $240 at 5% for 8 months. $ _____

10) $9,000 at 1.6% for 4 years. $ _____

✍ **Calculate each simple interest word problem.**

11) A new car, valued at $16,000, depreciates at 3.5% per year. What is the value of the car two year after purchase? $_____

12) Sara puts $9,000 into an investment yielding 8% annual simple interest; she left the money in for three years. How much interest does Sara get at the end of those three years? $_____

13) A bank is offering 12.5% simple interest on a savings account. If you deposit $32,400, how much interest will you earn in one years? $_____

14) $2,400 interest is earned on a principal of $10,000 at a simple interest rate of 12% interest per year. For how many years was the principal invested? _____

15) In how many years will $1,200 yield an interest of $384 at 8% simple interest? _____

16) Jim invested $5,000 in a bond at a yearly rate of 2.5%. He earned $375 in interest. How long was the money invested? _____

Answers of Worksheets – Chapter 3

Simplifying Ratios

1) 3:4
2) 1:10
3) 4:7
4) 1:3
5) 1:10
6) 1:8
7) 2:8
8) 2:5
9) 1:6
10) 7:9
11) 2:3
12) 7:2
13) 10:1

14) 3:4
15) 1:8
16) 5:7
17) 7:9
18) 3:1
19) 1:3
20) 7:1
21) 5:1
22) 5:4
23) 1:2
24) 1:30
25) $\frac{1}{2}$

26) $\frac{3}{5}$
27) $\frac{3}{7}$
28) $\frac{1}{3}$
29) $\frac{1}{3}$
30) $\frac{3}{14}$
31) $\frac{1}{2}$
32) $\frac{1}{5}$
33) $\frac{5}{12}$
34) $\frac{2}{9}$
35) $\frac{11}{13}$

36) $\frac{4}{9}$
37) $\frac{1}{12}$
38) $\frac{3}{11}$
39) $\frac{1}{15}$
40) $\frac{1}{4}$
41) $\frac{4}{3}$
42) $\frac{1}{5}$
43) $\frac{22}{41}$
44) $\frac{1}{9}$
45) $\frac{3}{5}$

Proportional Ratios

1) 12
2) 90
3) 5
4) 24
5) 63
6) 14
7) 23
8) 8
9) 48

10) 50
11) 14.2
12) 18
13) Yes
14) Yes
15) Yes
16) No
17) No
18) No

19) Yes
20) Yes
21) No
22) No
23) Yes
24) Yes
25) 24
26) 392
27) 10

28) 42
29) 48
30) 96
31) 96
32) 68
33) 8.4
34) 8
35) 73.5
36) 94

Similarity and ratios

1) 30
2) 2
3) 10

4) 11
5) 150 feet
6) 8.96 meters

7) 40 feet
8) 187.2 miles

Ratio and Rates Word Problems

1) 4:5

2) 252

3) The ratio for both classes is 9 to 5.

4) Walmart is a better buy.

5) 510, the rate is 30 per hour.

6) $39.20

7) 28

8) 18

Percentage Calculations

1) 1.8

2) 1.6

3) 20

4) 12

5) 27

6) 24.5

7) 4.2

8) 96

9) 43.2

10) 61

11) 182

12) 70.4

13) 88

14) 26.32

15) 25.5

16) 69.36

17) 72

18) 121

19) 50%

20) 80%

21) 5%

22) 25%

23) 12.5%

24) 25%

25) 25%

26) 3%

27) 5%

28) 14.33%

29) 30.30%

30) 54

Percent Problems

1) 50%

2) 40%

3) 160%

4) 4%

5) 1.5%

6) 6%

7) 20%

8) 8%

9) 30%

10) 25%

11) 50%

12) 50%

13) 8%

14) 125%

15) 150%

16) 30%

17) 50%

18) 320%

19) 75%

20) 14.4 ounces

21) 60

22) 75%

23) 20%

Discount, Tax and Tip

1) $486.00

2) $249.60

3) $1,008.00

4) $10,660.00

5) $412.00

6) $370,080

7) $144.00

8) $144.00

9) $56.00

10) $640.00

11) $64.40

12) $55.00

13) $117.50

14) $62.40

15) $80.00

16) $96.00

17) $43.35

18) $540.80

Percent of Change

1) 100%	7) 50%	13) 20%
2) 400%	8) 50%	14) 25%
3) 600%	9) 40%	15) 50%
4) 300%	10) 50%	16) 15%
5) 200%	11) 25%	17) 20%
6) 150%	12) 60%	

Simple Interest

1) $126.00	7) $1,536.00	13) $4,050.00
2) $216.00	8) $105.00	14) 2 years
3) $475.00	9) $9.80	15) 4 years
4) $56.875	10) $576.00	16) 3 years
5) $8.00	11) $14,880.00	
6) $5,880.00	12) $2,160.00	

Chapter 4:

Exponents and Radicals

Expressions

Topics that you will practice in this chapter:

- ✓ Multiplication Property of Exponents
- ✓ Zero and Negative Exponents
- ✓ Division Property of Exponents
- ✓ Powers of Products and Quotients
- ✓ Negative Exponents and Negative Bases
- ✓ Scientific Notation
- ✓ Square Roots
- ✓ Simplifying Radical Expressions
- ✓ Simplifying Radical Expressions Involving Fractions
- ✓ Multiplying Radical Expressions
- ✓ Adding and Subtracting Radical Expressions
- ✓ Domain and Range of Radical Functions
- ✓ Solving Radical Equations

Mathematics is no more computation than typing is literature.

– John Allen Paulos

Multiplication Property of Exponents

✎ **Simplify and write the answer in exponential form.**

1) $2 \times 2^5 =$

2) $7^2 \times 7 =$

3) $8^3 \times 8^3 =$

4) $9^4 \times 9^3 =$

5) $4^2 \times 4^4 \times 4 =$

6) $5 \times 5^2 \times 5^3 =$

7) $9^3 \times 9^3 \times 9 \times 9 =$

8) $4x \times x =$

9) $x^5 \times x^3 =$

10) $x^6 \times x^2 =$

11) $x^2 \times x^4 \times x^5 =$

12) $7x \times 7x =$

13) $4x^2 \times 5x^3 =$

14) $12x^3 \times x =$

15) $3x^2 \times 3x^2 \times 3x^2 =$

16) $7x^5 \times 2x^3 =$

17) $x^8 \times 2x =$

18) $3x \times 3x^3 =$

19) $6x^2 \times 2x^5 =$

20) $3yx^3 \times 12x =$

21) $8x^3 \times y^5x^2 =$

22) $4y^7x^2 \times 3y^2x^5 =$

23) $9yx^2 \times 4x^5y^2 =$

24) $10x^4 \times 11x^4y^4 =$

25) $9x^3y^4 \times 9x^6y^2 =$

26) $12x^4y^4 \times 6xy^3 =$

27) $9xy^4 \times 11x^3y^3 =$

28) $6x^2y^4 \times 8x^3y^6 =$

29) $8x \times y^7x^2 \times 5y^3 =$

30) $3yx^3 \times 2y^3x^2 \times 7xy =$

31) $8yx^5 \times 3y^4x \times 3xy^3 =$

32) $9x^3 \times 11y^4x^3 \times 2yx^4 =$

Zero and Negative Exponents

✍ **Evaluate the following expressions.**

1) $1^{-5} =$

2) $2^{-4} =$

3) $2^{-5} =$

4) $3^0 =$

5) $3^{-2} =$

6) $2^{-7} =$

7) $13^{-2} =$

8) $14^{-2} =$

9) $2^{-8} =$

10) $20^{-2} =$

11) $19^{-1} =$

12) $3^{-6} =$

13) $15^{-2} =$

14) $10^{-2} =$

15) $16^{-2=}$

16) $30^{-2} =$

17) $8^{-4} =$

18) $3^{-7} =$

19) $2^{-10} =$

20) $10^{-3} =$

21) $18^{-2} =$

22) $25^{-2} =$

23) $40^{-2} =$

24) $50^{-2} =$

25) $11^{-3} =$

26) $22^{-2} =$

27) $17^{-2} =$

28) $3^{-8} =$

29) $4^{-5} =$

30) $60^{-2} =$

31) $\left(\frac{1}{3}\right)^{-2}$

32) $\left(\frac{1}{5}\right)^{-3} =$

33) $\left(\frac{1}{8}\right)^{-2} =$

34) $\left(\frac{2}{5}\right)^{-2} =$

35) $\left(\frac{1}{15}\right)^{-2} =$

36) $\left(\frac{7}{12}\right)^{-2} =$

37) $\left(\frac{1}{20}\right)^{-2} =$

38) $\left(\frac{1}{7}\right)^{-3} =$

39) $\left(\frac{2}{3}\right)^{-5} =$

40) $\left(\frac{9}{11}\right)^{-1} =$

41) $\left(\frac{8}{9}\right)^{-2} =$

42) $\left(\frac{1}{8}\right)^{-3} =$

Division Property of Exponents

✍ **Simplify.**

1) $\dfrac{5^2}{5^6} =$

2) $\dfrac{6^9}{6^5} =$

3) $\dfrac{9^7}{9} =$

4) $\dfrac{3}{3^3} =$

5) $\dfrac{2x}{x^8} =$

6) $\dfrac{4 \times 4^5}{4^5 \times 4^2} =$

7) $\dfrac{12^6}{12^2} =$

8) $\dfrac{7 \times 7^9}{7^2 \times 7^4} =$

9) $\dfrac{4^5 \times 4^8}{4^2 \times 4^{11}} =$

10) $\dfrac{20x}{40x^4} =$

11) $\dfrac{8x^9}{9x^6} =$

12) $\dfrac{24x^3}{16x^5} =$

13) $\dfrac{25x^2}{50y^8} =$

14) $\dfrac{60xy^5}{12x^4y^2} =$

15) $\dfrac{8x^7}{12x} =$

16) $\dfrac{48x^2y^4}{16x^5} =$

17) $\dfrac{50x^6}{25x^9y^{14}} =$

18) $\dfrac{90yx^8}{15yx^9} =$

19) $\dfrac{18x^9y}{36x^{12}y^3} =$

20) $\dfrac{9x^8}{81x^8} =$

21) $\dfrac{9x^{-7}}{11x^{-3}} =$

Powers of Products and Quotients

✎ **Simplify.**

1) $(4^2)^3 =$

2) $(5^2)^2 =$

3) $(3 \times 3^2)^3 =$

4) $(3 \times 2^3)^2 =$

5) $(15^2 \times 15^2)^5 =$

6) $(7^2 \times 7^3)^4 =$

7) $(9 \times 9^2)^2 =$

8) $(4^6)^3 =$

9) $(7x^7)^3 =$

10) $(8x^4y^3)^2 =$

11) $(3x^3y^2)^4 =$

12) $(4x^2y^2)^2 =$

13) $(3x^5y^2)^3 =$

14) $(4x^3y^2)^3 =$

15) $(2x^3x)^5 =$

16) $(6x^4x^2)^2 =$

17) $(7x^{12}y^5)^2 =$

18) $(5x^7x^4)^3 =$

19) $(8x^2 \times 6x)^2 =$

20) $(9x^{14}y^3)^3 =$

21) $(5x^4y^2)^4 =$

22) $(3x^3y^7)^5 =$

23) $(8x \times 2y^3)^2 =$

24) $\left(\frac{8x}{x^3}\right)^3 =$

25) $\left(\frac{x^4y^5}{x^3y^5}\right)^7 =$

26) $\left(\frac{36xy}{6x^5}\right)^2 =$

27) $\left(\frac{x^4}{x^5y^2}\right)^3 =$

28) $\left(\frac{xy^2}{x^3y^8}\right)^{-3} =$

29) $\left(\frac{5xy^7}{x^2}\right)^3 =$

30) $\left(\frac{xy^5}{2xy^3}\right)^{-6} =$

Negative Exponents and Negative Bases

✎ **Simplify.**

1) $-4^{-2} =$

2) $-7^{-1} =$

3) $-5^{-2} =$

4) $-x^{-9} =$

5) $10x^{-2} =$

6) $-7x^{-4} =$

7) $-15x^{-4} =$

8) $-15x^{-7}y^{-4} =$

9) $32x^{-9}y^{-3} =$

10) $45a^{-7}b^{-3} =$

11) $-25x^3y^{-5} =$

12) $-\dfrac{18}{x^{-9}} =$

13) $-\dfrac{13x}{a^{-8}} =$

14) $\left(-\dfrac{1}{3}\right)^{-4} =$

15) $\left(-\dfrac{3}{4}\right)^{-3} =$

16) $-\dfrac{12}{a^{-6}b^{-4}} =$

17) $-\dfrac{48x}{x^{-6}} =$

18) $-\dfrac{a^{-12}}{b^{-5}} =$

19) $-\dfrac{27}{x^{-5}} =$

20) $\dfrac{12b}{-48c^{-6}} =$

21) $\dfrac{24ab}{a^{-4}b^{-3}} =$

22) $-\dfrac{8n^{-7}}{40p^{-9}} =$

23) $\dfrac{9ab^{-6}}{-5c^{-2}} =$

24) $\left(\dfrac{2a}{3c}\right)^{-4} =$

25) $\left(-\dfrac{8x}{5yz}\right)^{-2} =$

26) $\dfrac{9ab^{-6}}{-4c^{-3}} =$

27) $\left(-\dfrac{x^3}{x^4}\right)^{-5} =$

28) $\left(-\dfrac{x^{-3}}{3x^3}\right)^{-3} =$

29) $\left(-\dfrac{x^{-6}}{x^4}\right)^{-3} =$

Scientific Notation

✎ **Write each number in scientific notation.**

1) 0.226 =

2) 0.05 =

3) 4.8 =

4) 90 =

5) 120 =

6) 0.123 =

7) 82 =

8) 5,400 =

9) 2,460 =

10) 75,300 =

11) 61,000,000 =

12) 0.00009 =

13) 468,000 =

14) 0.00458 =

15) 0.000087 =

16) 31,800,000 =

17) 950,000 =

18) 9,000,000,000 =

19) 0.0007 =

20) 0.00041 =

✎ **Write each number in standard notation.**

21) 4×10^{-2} =

22) 7×10^{-4} =

23) 4.3×10^{6} =

24) 7×10^{-4} =

25) 8.7×10^{-3} =

26) 12×10^{5} =

27) 35×10^{3} =

28) 1.89×10^{5} =

29) 13×10^{-6} =

30) 7.3×10^{-4} =

Square Roots

✎ **Find the value each square root.**

1) $\sqrt{64} =$ ____

2) $\sqrt{4} =$ ____

3) $\sqrt{289} =$ ____

4) $\sqrt{0.25} =$ ____

5) $\sqrt{0.01} =$ ____

6) $\sqrt{0.09} =$ ____

7) $\sqrt{1,600} =$ ____

8) $\sqrt{2.25} =$ ____

9) $\sqrt{0} =$ ____

10) $\sqrt{0.04} =$ ____

11) $\sqrt{0.36} =$ ____

12) $\sqrt{0.81} =$ ____

13) $\sqrt{0.49} =$ ____

14) $\sqrt{1.21} =$ ____

15) $\sqrt{1.69} =$ ____

16) $\sqrt{0.16} =$ ____

17) $\sqrt{529} =$ ____

18) $\sqrt{625} =$ ____

19) $\sqrt{0.81} =$ ____

20) $\sqrt{20} =$ ____

21) $\sqrt{50} =$ ____

22) $\sqrt{676} =$ ____

23) $\sqrt{270} =$ ____

24) $\sqrt{32} =$ ____

✎ **Evaluate.**

25) $\sqrt{4} \times \sqrt{16} =$ _____

26) $\sqrt{49} \times \sqrt{64} =$ _____

27) $\sqrt{2} \times \sqrt{8} =$ _____

28) $\sqrt{17} \times \sqrt{17} =$ _____

29) $\sqrt{13} \times \sqrt{13} =$ _____

30) $\sqrt{15} \times \sqrt{15} =$ _____

31) $\sqrt{19} + \sqrt{19} =$ _____

32) $\sqrt{1} + \sqrt{1} =$ _____

33) $8\sqrt{7} - 2\sqrt{7} =$ _____

34) $7\sqrt{10} \times 6\sqrt{10} =$ _____

35) $9\sqrt{5} \times 2\sqrt{5} =$ _____

36) $8\sqrt{3} - \sqrt{12} =$ _____

Simplifying Radical Expressions

✍ **Simplify.**

1) $\sqrt{13y^2} =$

2) $\sqrt{60x^3} =$

3) $\sqrt[3]{27a} =$

4) $\sqrt{81x^2} =$

5) $\sqrt{150a} =$

6) $\sqrt[3]{135w^3} =$

7) $\sqrt{200x} =$

8) $\sqrt{192v} =$

9) $\sqrt[3]{64x} =$

10) $\sqrt{84x^3} =$

11) $\sqrt{121x^2} =$

12) $\sqrt[3]{48a} =$

13) $\sqrt{480} =$

14) $\sqrt{1,575p^2} =$

15) $\sqrt{108m^6} =$

16) $\sqrt{198x^3y^2} =$

17) $\sqrt{169x^2y^3} =$

18) $\sqrt{25a^6} =$

19) $\sqrt{50x^2y^3} =$

20) $\sqrt[3]{512y^3} =$

21) $2\sqrt{144x^2} =$

22) $3\sqrt{400x^2} =$

23) $\sqrt[3]{189xy^4} =$

24) $\sqrt[3]{1,331x^3y^5} =$

25) $3\sqrt{150a} =$

26) $\sqrt[3]{729y} =$

27) $3\sqrt{18xyr^3} =$

28) $6\sqrt{225x^2yz^6} =$

29) $3\sqrt[3]{125x^3y^2} =$

30) $7\sqrt{12a^2bc^4} =$

31) $4\sqrt[3]{1,000x^9y^{15}} =$

Answers of Worksheets – Chapter 4

Multiplication Property of Exponents

1) 2^6
2) 7^3
3) 8^6
4) 9^7
5) 4^7
6) 5^6
7) 9^8
8) $4x^2$

9) x^8
10) x^8
11) x^{11}
12) $49x^2$
13) $20x^5$
14) $12x^4$
15) $27x^6$
16) $14x^8$

17) $2x^8$
18) $9x^4$
19) $12x^7$
20) $36x^4y$
21) $8x^5y^5$
22) $12x^7y^9$
23) $36x^7y^3$
24) $110x^8y^4$

25) $81x^9y^6$
26) $72x^5y^7$
27) $99x^4y^7$
28) $48x^5y^{10}$
29) $40x^3y^{10}$
30) $42x^6y^5$
31) $72x^7y^8$
32) $198x^{10}y^5$

Zero and Negative Exponents

1) 1
2) $\frac{1}{16}$
3) $\frac{1}{32}$
4) 1
5) $\frac{1}{9}$
6) $\frac{1}{128}$
7) $\frac{1}{169}$
8) $\frac{1}{196}$
9) $\frac{1}{256}$
10) $\frac{1}{400}$
11) $\frac{1}{19}$

12) $\frac{1}{729}$
13) $\frac{1}{225}$
14) $\frac{1}{100}$
15) $\frac{1}{256}$
16) $\frac{1}{900}$
17) $\frac{1}{4,096}$
18) $\frac{1}{2,187}$
19) $\frac{1}{1,024}$
20) $\frac{1}{1,000}$
21) $\frac{1}{324}$

22) $\frac{1}{625}$
23) $\frac{1}{1,600}$
24) $\frac{1}{2,500}$
25) $\frac{1}{1,331}$
26) $\frac{1}{484}$
27) $\frac{1}{289}$
28) $\frac{1}{6,561}$
29) $\frac{1}{1,024}$
30) $\frac{1}{3,600}$
31)

32)
33)
34) 6.25
35) 225
36) $\frac{144}{49}$
37) 400
38) 343
39) $\frac{243}{32}$
40) $\frac{11}{9}$
41) $\frac{81}{64}$
42) 512

Division Property of Exponents

1) $\frac{1}{5^4}$
2) 6^4
3) 9^6
4) $\frac{1}{3^2}$

5) $\frac{2}{x^7}$
6) $\frac{1}{4}$
7) 12^4
8) 7^4
9) 1

10) $\frac{1}{2x^3}$
11) $\frac{8x^3}{9}$
12) $\frac{3}{2x^2}$
13) $\frac{x^2}{2y^8}$

14) $\frac{5y^3}{x^3}$
15) $\frac{2x^6}{3}$
16) $\frac{3y^4}{x^3}$
17) $\frac{2}{x^3y^{14}}$

18) $\frac{6}{x}$ 19) $\frac{1}{2x^3y^2}$ 20) $\frac{1}{9}$ 21) $\frac{9}{11x^4}$

Powers of Products and Quotients

1) 4^6
2) 5^4
3) 3^9
4) 24^2
5) 15^{20}
6) 7^{20}
7) 9^6
8) 4^{18}
9) $343x^{21}$
10) $64x^8y^6$
11) $81x^{12}y^8$

12) $16x^4y^4$
13) $27x^{15}y^6$
14) $64x^9y^6$
15) $32x^{20}$
16) $36x^{12}$
17) $49x^{24}y^{10}$
18) $125x^{33}$
19) $2,304x^6$
20) $729x^{42}y^9$
21) $625x^{16}y^9$
22) $243x^{15}y^8$

23) $256x^2y^6$
24) $\frac{512}{x^6}$
25) x^7
26) $\frac{36y^2}{x^8}$
27) $\frac{1}{x^3y^6}$
28) x^6y^{18}
29) $\frac{125y^{21}}{x^3}$
30) $\frac{64}{y^{12}}$

Negative Exponents and Negative Bases

1) $-\frac{1}{16}$
2) $-\frac{1}{7}$
3) $-\frac{1}{25}$
4) $-\frac{1}{x^9}$
5) $\frac{10}{x^2}$
6) $-\frac{7}{x^4}$
7) $-\frac{15}{x^4}$
8) $-\frac{15}{x^7y^4}$
9) $\frac{32}{x^9y^3}$
10) $\frac{45}{a^7b^3}$

11) $-\frac{25x^3}{y^5}$
12) $-18x^9$
13) $-13xa^8$
14) 81
15) $-\frac{64}{27}$
16) $-12a^6b^4$
17) $-48x^7$
18) $-\frac{b^5}{a^{12}}$
19) $-27x^5$
20) $-\frac{bc^6}{4}$
21) $24a^5b^4$

22) $-\frac{p^9}{5n^7}$
23) $-\frac{9ac^2}{5b^6}$
24) $\frac{81c^4}{16a^4}$
25) $\frac{25y^2z^2}{64x^2}$
26) $-\frac{9ac^3}{4b^6}$
27) $-x^5$
28) $-27x^{18}$
29) $-x^{30}$

Writing Scientific Notation

1) 2.26×10^{-1}
2) 5×10^{-2}
3) 4.8×10^0

4) 9×10^1
5) 1.2×10^2
6) 1.23×10^{-1}

7) 8.2×10^1
8) 5.4×10^3
9) 2.46×10^3

10) 7.53×10^4

11) 61×10^6

12) 9×10^{-5}

13) 4.68×10^5

14) 4.58×10^{-3}

15) 8.7×10^{-5}

16) 3.18×10^7

17) 9.5×10^5

18) 9×10^9

19) 7×10^{-4}

20) 4.1×10^{-4}

21) 0.04

22) 0.0007

23) $4,300,000$

24) 0.0007

25) 0.0087

26) $1,200,000$

27) $35,000$

28) $189,000$

29) 0.000013

30) 0.00073

Square Roots

1) 8

2) 2

3) 17

4) 0.5

5) 0.1

6) 0.3

7) 40

8) 1.5

9) 0

10) 0.2

11) 0.6

12) 0.9

13) 0.7

14) 1.1

15) 1.3

16) 0.4

17) 23

18) 25

19) 0.9

20) $2\sqrt{5}$

21) $5\sqrt{2}$

22) 26

23) $3\sqrt{30}$

24) $4\sqrt{2}$

25) 8

26) 56

27) 4

28) 17

29) 13

30) 15

31) $2\sqrt{19}$

32) 2

33) $6\sqrt{7}$

34) 420

35) 90

36) $6\sqrt{3}$

Simplifying radical expressions

1) $y\sqrt{13}$

2) $2x\sqrt{15x}$

3) $3\sqrt[3]{a}$

4) $9x$

5) $5\sqrt{6a}$

6) $3w\sqrt[3]{5}$

7) $10\sqrt{2x}$

8) $8\sqrt{3v}$

9) $4\sqrt[3]{x}$

10) $2x\sqrt{21x}$

11) $11x$

12) $2\sqrt[3]{6a}$

13) $4\sqrt{30}$

14) $15p\sqrt{7}$

15) $6m^3\sqrt{3}$

16) $3x.y\sqrt{22x}$

17) $13xy\sqrt{y}$

18) $5a^3$

19) $5xy\sqrt{2y}$

20) $8y$

21) $24x$

22) $60x$

23) $3y\sqrt[3]{7xy}$

24) $11xy\sqrt[3]{y^2}$

25) $15\sqrt{6a}$

26) $9\sqrt[3]{y}$

27) $9r\sqrt{2xyr}$

28) $90xz^3\sqrt{y}$

29) $15x\sqrt[3]{y^2}$

30) $14ac^2\sqrt{b}$

31) $40x^3y^{15}$

Chapter 5:

Algebraic Expressions

Topics that you will practice in this chapter:

- ✓ Simplifying Variable Expressions
- ✓ Simplifying Polynomial Expressions
- ✓ Translate Phrases into an Algebraic Statement
- ✓ The Distributive Property
- ✓ Evaluating One Variable Expressions
- ✓ Evaluating Two Variables Expressions
- ✓ Combining like Terms

Mathematics is, as it were, a sensuous logic, and relates to philosophy as do the arts, music, and plastic art to poetry. — *K. Shegel*

Simplifying Variable Expressions

✎ **Simplify each expression.**

1) $3(x + 8) =$

2) $(-4)(7x - 3) =$

3) $11x + 8 - 7x =$

4) $-6 - 2x^2 - 9x^2 =$

5) $8 + 17x^2 + 6 =$

6) $9x^2 + 13x + 19x^2 =$

7) $7x^2 - 15x^2 + 3x =$

8) $8x^2 - 11x - 3x =$

9) $3x + 9(1 - 4x) =$

10) $14x + 2(20x - 4) =$

11) $6(-3x - 7) - 22 =$

12) $7x^2 + (-12x) =$

13) $x - 8 + 15 - 7x =$

14) $3 - 6x + 12 - 3x =$

15) $20x - 14 + 27 + 12x =$

16) $(-7)(6x - 5) + 14x =$

17) $11x - 4(3 - 7x) =$

18) $22x + 3(5x + 2) + 11 =$

19) $4(-3x + 8) + 10x =$

20) $16x - 3x(2x + 7) =$

21) $9x + 12x(2 - 4x) =$

22) $5x(-4x + 11) + 18x =$

23) $20x + 24x + 3x^2 =$

24) $7x(x - 7) - 28 =$

25) $7x - 12 + 5x + 3x^2 =$

26) $4x^2 - 9x - 12x =$

27) $8x - 22x^2 - 21x^2 - 11 =$

28) $9 + 3x^2 - 8x^2 - 27x =$

29) $14x + 2x^2 + 4x + 28 =$

30) $7x^2 + 45x + 12x^2 =$

31) $25 + 15x^2 + 9x - 3x^2 =$

32) $17x - 32x - 2x^2 + 30 =$

Simplifying Polynomial Expressions

✎ **Simplify each polynomial.**

1) $(5x^4 + 2x^2) - (11x + 6x^2) =$ _____

2) $(x^7 + 6x^4) - (8x^4 + 4x^2) =$ _____

3) $(24x^5 + 8x^3) - (x^3 - 12x^5) =$ _____

4) $14x - 9x^5 - 4(7x^5 + 7x^3) =$ _____

5) $(7x^4 - 5) + 2(4x^2 - 8x^4) =$ _____

6) $(9x^5 - 3x) - 3(8x^5 - 3x^4) =$ _____

7) $4(2x - 4x^4) - 5(3x^4 + x^2) =$ _____

8) $(4x^2 - 2x) - (5x^3 + 9x^2) =$ _____

9) $8x^4 - (9x^6 + 2x) + 2x^2 =$ _____

10) $x^5 - 3(x^3 + 2x) + 9x =$ _____

11) $(4x^2 - 2x^5) - (4x^5 - 2x^2) =$ _____

12) $8x^3 - 8x^5 + 17x^4 - 12x^5 =$ _____

13) $4x^3 - 9x^7 + 18x^7 - 24x^6 =$ _____

14) $4x^5 + 13x^3 - 17x^5 + 24x =$ _____

15) $7x^6 - 9x^7 + 5x^6 - 12x^3 =$ _____

16) $4x^4 + 19x - 3x^3 - 21x^4 =$ _____

Translate Phrases into an Algebraic Statement

✎ **Write an algebraic expression for each phrase.**

1) 13 multiplied by x. _____

2) Subtract 15 from y. _____

3) 22 divided by x. _____

4) 27 decreased by y. _____

5) Add y to 31. _____

6) The square of 7. _____

7) x raised to the seventh power. _____

8) The sum of five and a number. _____

9) The difference between forty–nine and y. _____

10) The quotient of eight and a number. _____

11) The quotient of the square of x and 34. _____

12) The difference between x and 14 is 41. _____

13) 7 times b reduced by the square of a. _____

14) Subtract the product of a and b from 51. _____

The Distributive Property

✎ **Use the distributive property to simply each expression.**

1) $4(2 + 5x) =$

2) $5(2 + 4x) =$

3) $6(5x - 5) =$

4) $(6x - 3)(-7) =$

5) $(-4)(x + 8) =$

6) $(4 + 4x)6 =$

7) $(-5)(8 - 7x) =$

8) $-(-3 - 12x) =$

9) $(-8x + 3)(-5) =$

10) $(-5)(x - 11) =$

11) $-(8 - 2x) =$

12) $3(7 + 4x) =$

13) $4(8 + 3x) =$

14) $(-8x + 2)5 =$

15) $(4 - 7x)(-9) =$

16) $(-12)(3x + 5) =$

17) $(9 - 3x)5 =$

18) $4(4 + 7x) =$

19) $12(3x - 6) =$

20) $(-7x + 5)4 =$

21) $(4 - 9x)(-2) =$

22) $(-15)(2x - 3) =$

23) $(14 - 3x)3 =$

24) $(-5)(10x - 4) =$

25) $(5 - 7x)(-12) =$

26) $(-8)(2x + 9) =$

27) $(-5 + 8x)(-7) =$

28) $(-6)(2 - 15x) =$

29) $13(4x - 6) =$

30) $(-15x + 13)(-4) =$

31) $(-9)(3x - 2) + 2(x + 5) =$

32) $(-9)(2x + 2) - (7 + 4x) =$

Evaluating One Variable Expressions

✎ **Evaluate each expression using the value given.**

1) $8 - x , x = 5$

2) $x - 10, x = 6$

3) $3x - 6, x = 5$

4) $x - 15, x = -2$

5) $12 - x , x = 4$

6) $x + 7, x = 1$

7) $2x + 9, x = 7$

8) $x + (-4), x = -7$

9) $2x + 9, x = 4$

10) $3x + 10, x = -2$

11) $18 + 2x - 4, x = -1$

12) $18 - 6x, x = 2$

13) $8x - 2, x = 4$

14) $2x - 17, x = 8$

15) $13x - 12, x = 3$

16) $8 - 5x, x = -2$

17) $3(5x + 4), x = 5$

18) $4(-2x - 7), x = 3$

19) $7x - 5x + 12, x = 2$

20) $(8x + 4) \div 2, x = 6$

21) $(x + 15) \div 4, x = 9$

22) $6x - 10 + 3x, x = -5$

23) $(7 - 4x)(-3), x = -3$

24) $12x^2 + 5x - 4, x = 2$

25) $x^2 - 15x, x = -4$

26) $3x(3 - 6x), x = 2$

27) $13x + 8 - 6x^2, x = -3$

28) $(-2)(15x - 11 + 4x), x = 4$

29) $(-6) + \frac{x}{6} + x, x = 18$

30) $(-9) + \frac{x}{4}, x = 32$

31) $\left(-\frac{45}{x}\right) - 5 + 2x, x = 9$

32) $\left(-\frac{36}{x}\right) - 9 + 3x, x = 3$

Evaluating Two Variables Expressions

✎ **Evaluate each expression using the values given.**

1) $5x - y$,

   $x = 4, y = 3$

2) $3x + 2y$,

   $x = -2, y = 2$

3) $-6a + 5b$,

   $a = 3, b = 1$

4) $3x + 7 - y$,

   $x = 8, y = 4$

5) $5z + 12 - 3k$,

   $z = 5, k = 2$

6) $6(-x - 3y)$,

   $x = 5, y = 4$

7) $7a + 4b$,

   $a = 3, b = 5$

8) $8x \div 4y$,

   $x = 6, y = 4$

9) $2x + 18 + 4y$,

   $x = -3, y = 3$

10) $5a - (18 - 2b)$,

    $a = 5, b = 8$

11) $6z + 12 + 3k$,

    $z = -3, k = 3$

12) $2xy + 6 + 7x$,

    $x = 5, y = 3$

13) $6x + 2y - 9 + 3$,

    $x = 3, y = 2$

14) $\left(-\frac{21}{x}\right) + 6 + 3y$,

    $x = 7, y = 4$

15) $(-4)(-3a - b)$,

    $a = 2, b = 6$

16) $18 + 4x + 9 - 5y$,

    $x = 6, y = 4$

17) $7x + 5 - 6y + 11$,

    $x = 9, y = 3$

18) $9 + 4(-5x - 3y)$,

    $x = 4, y = 5$

19) $3x + 15 + 6y$,

    $x = 2, y = 4$

20) $7a - (4a - 2b) + 8$,

    $a = 3, b = 1$

Combining like Terms

✎ **Simplify each expression.**

1) $7x + 2x + 8 =$

2) $3(6x - 2) =$

3) $10x - 12x + 8 =$

4) $20x - 32x + 14 =$

5) $16x - 6x - 12 =$

6) $18x - 21 + 4x =$

7) $15 - (3x + 9) =$

8) $-14x + 7 - 11x =$

9) $5x - 10 - 3x + 1 =$

10) $24x + 7x - 22 =$

11) $14x + 8x - 2 =$

12) $(-4x + 2)8 =$

13) $34 + 6x + 8x - 4 =$

14) $3(x - 8x) - 5 =$

15) $4(2x + 7) + 5x =$

16) $x - 27 - 9x =$

17) $3(5 + 4x) - 8x =$

18) $41x + 24 + 3x =$

19) $(-8x) + 30 + 15x =$

20) $(-4x) - 12 + 19x =$

21) $5(2x + 6) + 9x =$

22) $3(6 - 7x) - 11x =$

23) $-8x - (16 - 14x) =$

24) $(-9) - (6)(5x + 9) =$

25) $(-4)(6x - 5) - 12x =$

26) $-34x + 14 + 9x - 21x =$

27) $5(-13x + 6) - 24x =$

28) $-7x - 20 + 15x =$

29) $42x - 31x + 15 - 12x =$

30) $4(8x + 5x) - 17 =$

31) $54 - 22x - 28 - 19x =$

32) $-9(-7x - 11x) + 58x =$

Answers of Worksheets – Chapter 5

Simplifying Variable Expressions

1) $3x + 24$
2) $-28x + 12$
3) $4x + 8$
4) $-11x^2 - 6$
5) $17x^2 + 14$
6) $28x^2 + 13x$
7) $-8x^2 + 3x$
8) $8x^2 - 14x$
9) $-33x + 9$
10) $54x - 8$
11) $-18x - 64$

12) $7x^2 - 12x$
13) $-6x + 7$
14) $-9x + 15$
15) $32x + 13$
16) $-28x + 35$
17) $39x - 12$
18) $37x + 17$
19) $-2x + 32$
20) $-6x^2 - 5x$
21) $-48x^2 + 33x$
22) $-20x^2 + 73x$

23) $3x^2 + 44x$
24) $7x^2 - 49x - 28$
25) $3x^2 + 12x - 12$
26) $4x^2 - 21x$
27) $-43x^2 + 8x - 11$
28) $-5x^2 - 27x + 9$
29) $2x^2 + 18x + 28$
30) $19x^2 + 45x$
31) $12x^2 + 9x + 25$
32) $-2x^2 - 15x + 30$

Simplifying Polynomial Expressions

1) $5x^4 - 4x^2 - 11x$
2) $x^7 - 2x^4 - 4x^2$
3) $36x^5 + 7x^3$
4) $-37x^5 - 28x^2 + 14x$
5) $-9x^4 + 8x^2 - 5$
6) $-15x^5 + 9x^4 - 3x$
7) $-31x^3 - 5x^2 + 8x$
8) $-5x^3 - 5x^2 - 2x$

9) $-9x^6 + 8x^4 + 2x^2 - 2x$
10) $x^5 - 3x^3 + 3x$
11) $-6x^5 + 6x^2$
12) $-20x^5 + 17x^4 + 8x^3$
13) $9x^7 - 24x^6 + 4x^3$
14) $-13x^5 + 13x^3 + 24x$
15) $-9x^7 + 12x^6 - 12x^3$
16) $-17x^4 - 3x^3 + 19x$

Translate Phrases into an Algebraic Statement

1) $13x$
2) $y - 15$
3) $\frac{22}{x}$
4) $27 - y$

5) $y + 31$
6) 7^2
7) x^7
8) $5 + x$

9) $49 - y$
10) $\frac{8}{x}$
11) $\frac{x^2}{34}$

12) $x - 14 = 41$
13) $7b - a^2$
14) $51 - ab$

The Distributive Property

1) $20x + 8$
2) $20x + 10$

3) $30x - 30$
4) $-42x + 21$

5) $-4x - 32$
6) $24x + 24$

7) $35x - 40$
8) $12x + 3$

9) $40x - 15$

10) $-5x + 55$

11) $2x - 8$

12) $12x + 21$

13) $12x + 32$

14) $-40x + 10$

15) $63x - 36$

16) $-36x - 60$

17) $-15x + 45$

18) $28x + 16$

19) $36x - 72$

20) $-28x + 20$

21) $18x - 8$

22) $-30x + 45$

23) $-9x + 42$

24) $-50x + 20$

25) $84x - 60$

26) $-16x - 72$

27) $56x - 35$

28) $90x - 12$

29) $52x - 78$

30) $60x - 52$

31) $-25x + 28$

32) $-22x - 25$

Evaluating One Variables

1) 3

2) −4

3) 9

4) −17

5) 8

6) 8

7) 23

8) − 11

9) 17

10) 4

11) 12

12) 6

13) 30

14) −1

15) 27

16) 18

17) 87

18) −52

19) 16

20) 26

21) 6

22) −55

23) −57

24) 54

25) 76

26) −54

27) −85

28) −130

29) 15

30) −1

31) 8

32) −12

Evaluating Two Variables

1) 17

2) −2

3) −13

4) 27

5) 31

6) −102

7) 41

8) 3

9) 24

10) 23

11) 3

12) 71

13) 16

14) 15

15) 48

16) 31

17) 61

18) −131

19) 45

20) 19

Combining like Terms

1) $9x + 8$

2) $18x - 6$

3) $-2x + 8$

4) $-12x + 14$

5) $10x - 12$

6) $22x - 22$

7) $-3x + 6$

8) $-25x + 7$

9) $2x - 9$

10) $31x - 22$

11) $22x - 2$

12) $-32x + 16$

13) $14x + 30$

14) $-21x - 5$

15) $13x + 28$

16) $-8x - 27$

17) $4x + 15$

18) $44x + 24$

19) $7x + 30$

20) $15x - 12$

21) $19x + 30$

22) $-32x + 18$

23) $6x - 16$

24) $-30x - 63$

25) $-36x + 20$

26) $-46x + 14$

27) $-89x + 30$

28) $8x - 20$

29) $-x + 15$

30) $52x - 17$

31) $-41x + 26$

32) $220x$

Chapter 6:

Equations and Inequalities

Topics that you will practice in this chapter:

- ✓ One–Step Equations
- ✓ Multi–Step Equations
- ✓ Graphing Single–Variable Inequalities
- ✓ One–Step Inequalities
- ✓ Multi-Step Inequalities
- ✓ Systems of Equations
- ✓ Systems of Equations Word Problems

"Life is a math equation. In order to gain the most, you have to know how to convert negatives into positives." – Anonymous

One–Step Equations

✎ **Find the answer for each equation.**

1) $3x = 90, x =$ ____

2) $5x = 35, x =$ ____

3) $9x = 36, x =$ ____

4) $25x = 150, x =$ ____

5) $x + 18 = 23, x =$ ____

6) $x - 3 = 8, x =$ ____

7) $x - 7 = 4, x =$ ____

8) $x + 22 = 30, x =$ ____

9) $x - 11 = 6, x =$ ____

10) $24 = 28 + x, x =$ ____

11) $x - 5 = 7, x =$ ____

12) $9 - x = -7, x =$ ____

13) $43 = -8 + x, x =$ ____

14) $x - 23 = -38, x =$ ____

15) $x + 45 = -27, x =$ ____

16) $42 = 56 - x, x =$ ____

17) $-18 + x = -32, x =$ ____

18) $x - 13 = 7, x =$ ____

19) $35 = x - 10, x =$ ____

20) $x - 8 = -21, x =$ ____

21) $x - 54 = -20, x =$ ____

22) $x - 42 = -47, x =$ ____

23) $x - 8 = 29, x =$ ____

24) $-93 = x - 51, x =$ ____

25) $x + 15 = 37, x =$ ____

26) $108 = 12x, x =$ ____

27) $x - 33 = 27, x =$ ____

28) $x - 12 = 23, x =$ ____

29) $72 - x = 18, x =$ ____

30) $x + 34 = 58, x =$ ____

31) $21 - x = -9, x =$ __

32) $x - 59 = -80, x =$ ____

Multi–Step Equations

✎ **Find the answer for each equation.**

1) $3x + 1 = 7$

2) $-x + 10 = 9$

3) $5x - 13 = 7$

4) $-(4 - x) = 5$

5) $3x - 8 = 16$

6) $15x - 13 = 17$

7) $3x - 28 = 2$

8) $9x + 21 = 39$

9) $14x + 17 = 45$

10) $-14(8 + x) = 70$

11) $8(10 + x) = 32$

12) $16 = -(x - 8)$

13) $5(7 - 3x) = 50$

14) $-19 = -(3x + 7)$

15) $30(3 + x) = 60$

16) $9(x - 12) = 54$

17) $-24 = 3x + 5x$

18) $5x + 28 = -2x - 7$

19) $9(5 + 4x) = -99$

20) $18 - x = -12 - 6x$

21) $4 - 4x = 28 - 2x$

22) $15 + 12x = -15 + 8x$

23) $54 = (-3x) - 8 + 8$

24) $12 = 7x - 18 + 5x$

25) $-18 = -9x - 42 + 5x$

26) $11x - 6 = -33 + 8x$

27) $8x - 42 = 3x + 3$

28) $-15 - 8x = 4(5 - x)$

29) $x - 9 = -5(9 - 2x)$

30) $14x - 65 = -x - 110$

31) $3x - 129 = -3(11 + 7x)$

32) $-7x - 20 = 2x + 43$

Graphing Single–Variable Inequalities

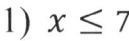 **Draw a graph for each inequality.**

1) $x \leq 7$

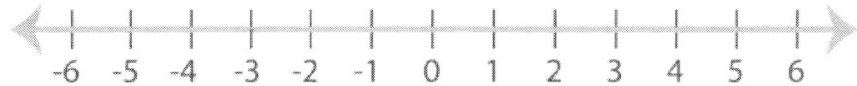

2) $x \leq -1.5$

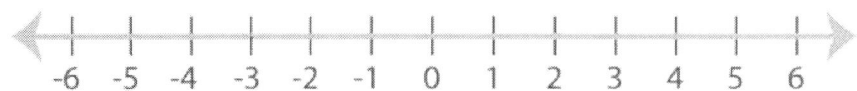

3) $x < -4$

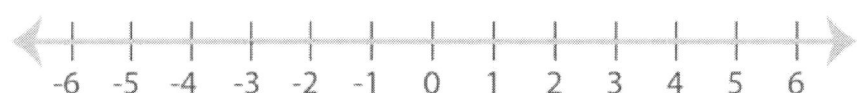

4) $x > 2.5$

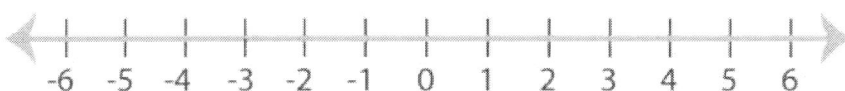

5) $x > 1.3$

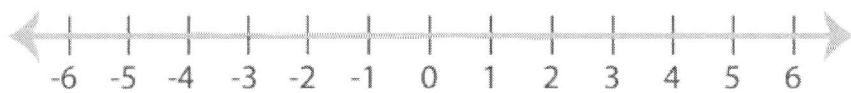

6) $x < 4$

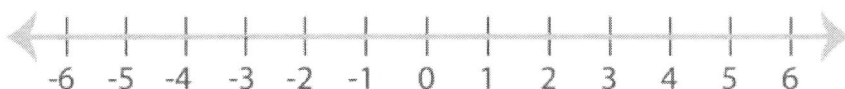

7) $x < 2.4$

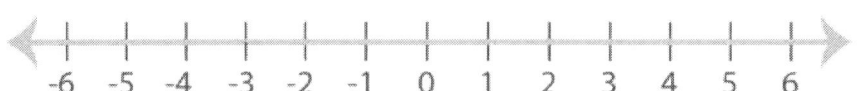

8) $x > -\frac{18}{10}$

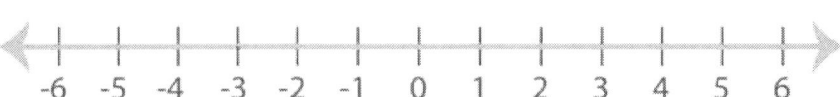

One–Step Inequalities

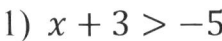 **Find the answer for each inequality and graph it.**

1) $x + 3 > -5$

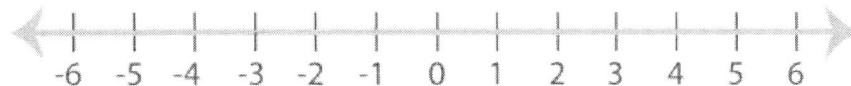

2) $x - 4 < 1$

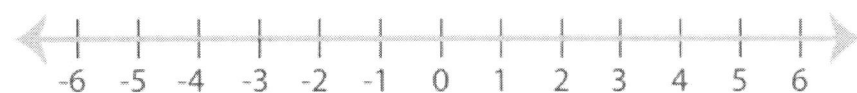

3) $7x < 42$

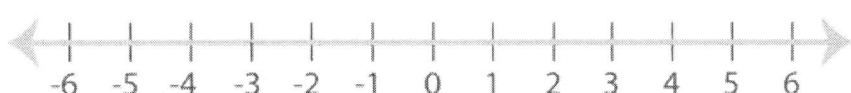

4) $13 + x > 12$

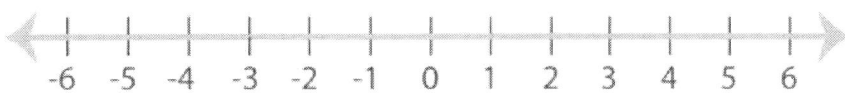

5) $x + 20 < 13$

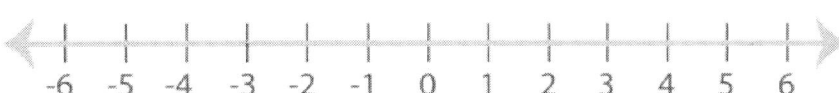

6) $14x \leq 42$

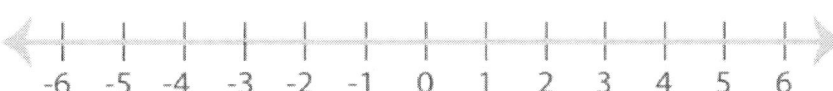

7) $11x \leq -44$

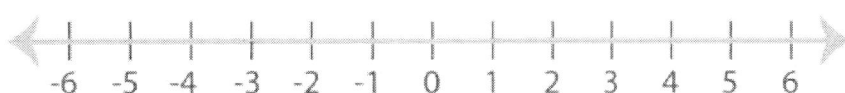

8) $x + 26 > 35$

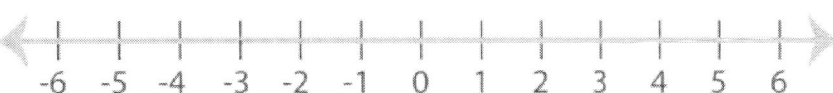

Multi-Step Inequalities

✎ **Calculate each inequality.**

1) $x - 8 \leq 12$

2) $9 - 3x \leq 18$

3) $4x - 7 \leq 9$

4) $8x - 9 \geq 15$

5) $x - 19 \geq 24$

6) $5x - 15 \leq 40$

7) $7x - 4 \leq 24$

8) $-18 + 8x \leq 22$

9) $9(x - 8) \leq 27$

10) $4x - 8 \leq 16$

11) $11x - 42 < 22$

12) $10x - 18 < 52$

13) $17 - 9x \geq -46$

14) $32 + 2x < 68$

15) $8 + 8x \geq 80$

16) $11 + 6x < 65$

17) $9x - 13 < 23$

18) $8(12 - 4x) \geq -68$

19) $-(2 + 5x) < 42$

20) $14 - 9x \geq -31$

21) $-5(x - 3) > 65$

22) $\dfrac{2x + 8}{3} \leq 12$

23) $\dfrac{8x + 16}{4} \leq 24$

24) $\dfrac{2x - 22}{9} > 8$

25) $7 + \dfrac{x}{4} < 21$

26) $\dfrac{32x}{16} - 4 < 6$

27) $\dfrac{12x + 36}{22} > 3$

28) $42 + \dfrac{x}{3} < 15$

Systems of Equations

✎ **Calculate each system of equations.**

1) $-6x + 7y = 8$ $x =$ _____
   $x + 4y = 9$ $y =$ _____

2) $-4x + 12y = 12$ $x =$ _____
   $14x - 16y = 10$ $y =$ _____

3) $y = -9$ $x =$ _____
   $2x - 5y = 12$ $y =$ _____

4) $4y = -4x + 20$ $x =$ _____
   $8x - 2y = -12$ $y =$ _____

5) $10x - 9y = -13$ $x =$ _____
   $-5x + 3y = 11$ $y =$ _____

6) $-6x - 8y = 10$ $x =$ _____
   $4x - 8y = 20$ $y =$ _____

7) $5x - 14y = -23$ $x =$ _____
   $-6x + 7y = 8$ $y =$ _____

8) $-4x + 3y = 3$ $x =$ _____
   $-x + 2y = 5$ $y =$ _____

9) $-4x + 5y = 15$ $x =$ _____
   $-3x + 4y = -10$ $y =$ _____

10) $-6x - 6y = -21$ $x =$ _____
    $-6x + 6y = -66$ $y =$ _____

11) $12x - 21y = 6$ $x =$ _____
    $-6x - 3y = -12$ $y =$ _____

12) $-4x - 4y = -14$ $x =$ _____
    $4x - 4y = 44$ $y =$ _____

13) $4x + 5y = 3$ $x =$ _____
    $3x - y = 6$ $y =$ _____

14) $3x - 2y = 2$ $x =$ _____
    $10x - 10y = 20$ $y =$ _____

15) $5x + 8y = 14$ $x =$ _____
    $-3x - 2y = -3$ $y =$ _____

16) $8x + 5y = 4$ $x =$ _____
    $-3x - 4y = 15$ $y =$ _____

Systems of Equations Word Problems

✍ **Find the answer for each word problem.**

1) Tickets to a movie cost $6 for adults and $4 for students. A group of friends purchased 9 tickets for $50.00. How many adults ticket did they buy? ____

2) At a store, Eva bought two shirts and five hats for $77.00. Nicole bought three same shirts and four same hats for $84.00. What is the price of each shirt? _____

3) A farmhouse shelters 10 animals, some are pigs, and some are ducks. Altogether there are 36 legs. How many pigs are there? _____

4) A class of 85 students went on a field trip. They took 24 vehicles, some cars and some buses. If each car holds 3 students and each bus hold 16 students, how many buses did they take? _____

5) A theater is selling tickets for a performance. Mr. Smith purchased 8 senior tickets and 10 child tickets for $248 for his friends and family. Mr. Jackson purchased 4 senior tickets and 6 child tickets for $132. What is the price of a senior ticket? $_____

6) The difference of two numbers is 15. Their sum is 33. What is the bigger number? $_____

7) The sum of the digits of a certain two–digit number is 7. Reversing its digits increase the number by 9. What is the number? _____

8) The difference of two numbers is 11. Their sum is 25. What are the numbers? _____

9) The length of a rectangle is 5 meters greater than 2 times the width. The perimeter of rectangle is 28 meters. What is the length of the rectangle? _____

10) Jim has 23 nickels and dimes totaling $2.40. How many nickels does he have? _____

Answers of Worksheets – Chapter 6

One–Step Equations

1) 30	9) 17	17) -14	25) 22
2) 7	10) -4	18) 20	26) 9
3) 4	11) 12	19) 45	27) 60
4) 6	12) 16	20) -13	28) 35
5) 5	13) 51	21) 34	29) 54
6) 11	14) -15	22) -5	30) 24
7) 11	15) -72	23) 37	31) 30
8) 8	16) 14	24) -42	32) -21

Multi–Step Equations

1) 2	9) 2	17) -3	25) -6
2) 1	10) -13	18) -5	26) -9
3) 4	11) -6	19) -4	27) 9
4) 9	12) -8	20) -6	28) -8.75
5) 8	13) -1	21) -12	29) 4
6) 2	14) 4	22) -7.5	30) -3
7) 10	15) -1	23) -18	31) 4
8) 2	16) 18	24) 2.5	32) -7

Graphing Single–Variable Inequalities

1)

2)

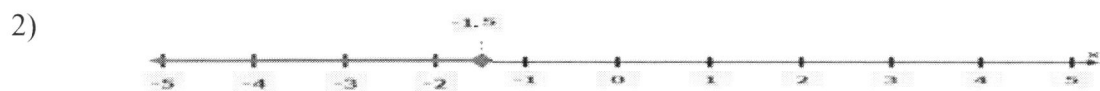

3)

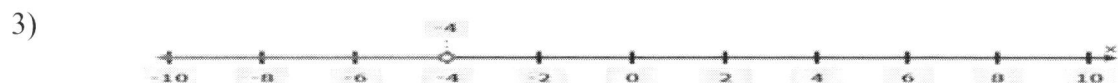

4)

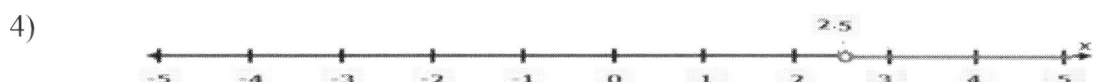

5)

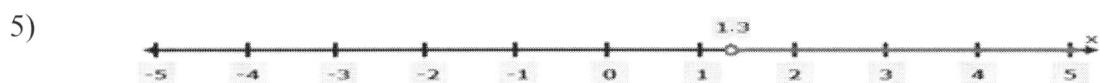

6)

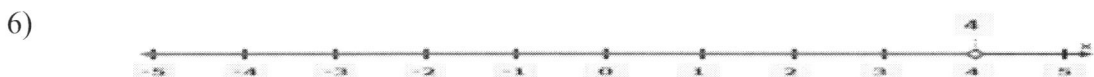

7)

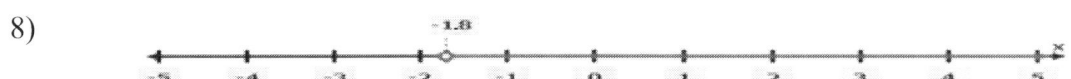

8)

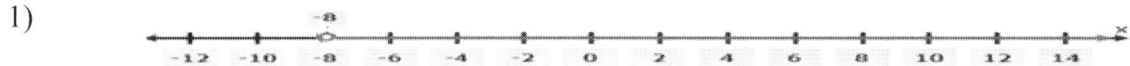

One–Step Inequalities

1)

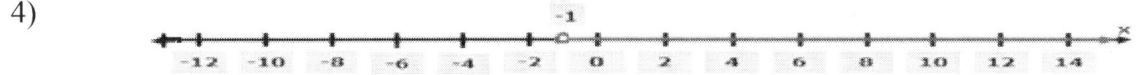

2)

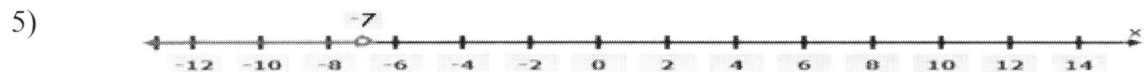

3)

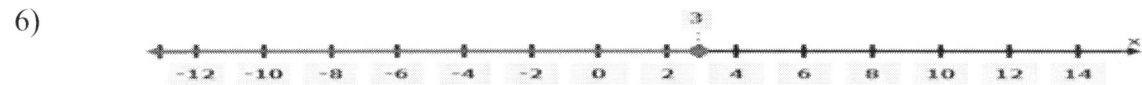

4)

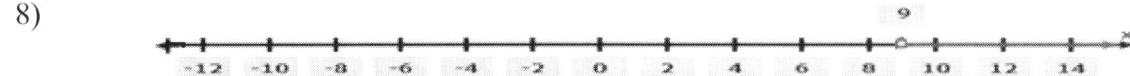

5)

6)

7)

8)

Multi-Step Inequalities

1) $x \leq 20$	5) $x \geq 43$	9) $x \leq 11$	13) $x \leq 7$
2) $x \geq -3$	6) $x \leq 11$	10) $x \leq 6$	14) $x < 18$
3) $x \leq 4$	7) $x \leq 4$	11) $x < 64/11$	15) $x \geq 9$
4) $x \geq 3$	8) $x \leq 5$	12) $x < 7$	16) $x < 9$

17) $x < 4$

18) $x \leq 41/8$

19) $x > -44/5$

20) $x \leq 5$

21) $x < -10$

22) $x \leq 14$

23) $x \leq 10$

24) $x > 47$

25) $x < 56$

26) $x < 9/4$

27) $x > 2.5$

28) $x < -81$

Systems of Equations

1) $x = 1, y = 2$

2) $x = 3, y = 2$

3) $x = -\dfrac{33}{2}$

4) $x = -\dfrac{1}{5}, y = \dfrac{26}{5}$

5) $x = -4, y = -3$

6) $x = 1, y = -2$

7) $x = 1, y = 2$

8) $x = \dfrac{9}{5}, y = \dfrac{17}{5}$

9) $x = -110, y = -85$

10) $x = -\dfrac{15}{4}, y = \dfrac{29}{4}$

11) $x = \dfrac{5}{3}, y = \dfrac{2}{3}$

12) $x = -\dfrac{15}{4}, y = \dfrac{29}{4}$

13) $x = \dfrac{33}{19}, y = -\dfrac{15}{19}$

14) $x = -2, y = -4$

15) $x = -\dfrac{2}{7}, y = \dfrac{27}{14}$

16) $x = \dfrac{91}{17}, y = -\dfrac{132}{17}$

Systems of Equations Word Problems

1) 7

2) $16

3) 8

4) 1

5) $21

6) 24

7) 43

8) 18, 7

9) 11 meters

10) 18

Chapter 7:

Linear Functions

Topics that you will practice in this chapter:

- ✓ Finding Slope
- ✓ Graphing Lines Using Line Equation
- ✓ Writing Linear Equations
- ✓ Graphing Linear Inequalities
- ✓ Finding Midpoint
- ✓ Finding Distance of Two Points

"Nature is written in mathematical language." – Galileo Galilei

Finding Slope

✎ **Find the slope of each line.**

1) $y = 2x + 5$

2) $y = -x + 17$

3) $y = 4x + 16$

4) $y = -3x + 15$

5) $y = 27 + 7x$

6) $y = 11 - 4x$

7) $y = 7x + 14$

8) $y = -8x + 18$

9) $y = -9x + 15$

10) $y = 8x - 13$

11) $y = \frac{1}{5}x + 9$

12) $y = -\frac{3}{7}x + 19$

13) $-3x + 6y = 17$

14) $4x + 4y = 16$

15) $8y - 3x = 32$

16) $11y - 3x = 42$

✎ **Find the slope of the line through each pair of points.**

17) $(1, 8), (5, 16)$

18) $(-2, 14), (2, 18)$

19) $(7, -1), (3, 9)$

20) $(-4, -4), (2, 14)$

21) $(16, -1), (4, 11)$

22) $(-21, 5), (-10, 38)$

23) $(8, 11), (12, 19)$

24) $(22, -22), (10, 14)$

25) $(21, -15), (19, -13)$

26) $(11, 10), (7, -2)$

27) $(5, 4), (9, 16)$

28) $(34, -87), (22, 45)$

Graphing Lines Using Line Equation

 Sketch the graph of each line.

1) $y = x - 5$

2) $y = -3x + 4$

3) $x - 2y = 0$

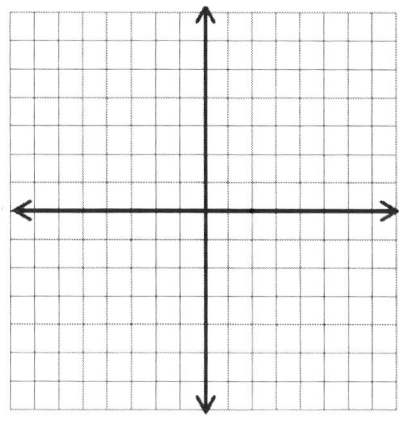

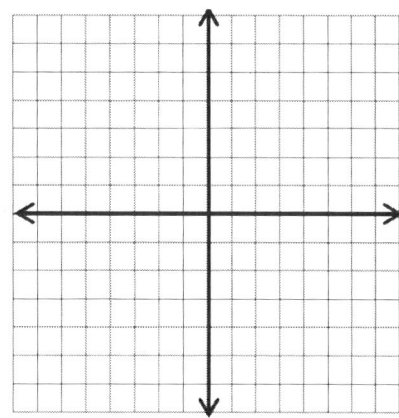

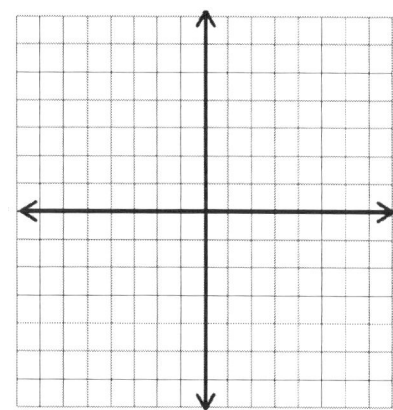

4) $x + y = -4$

5) $4x + 3y = -2$

6) $y - 3x + 6 = 0$

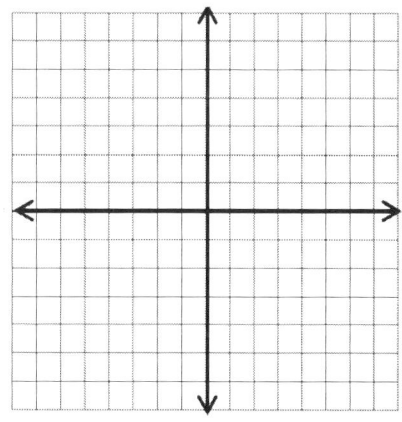

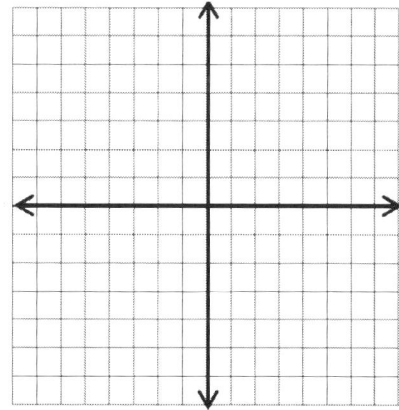

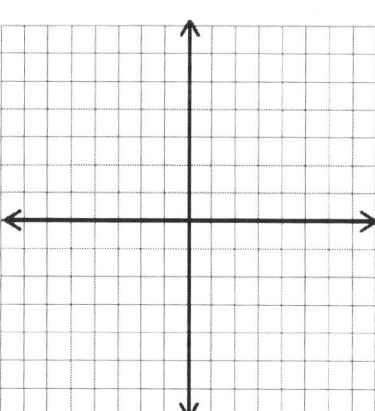

Writing Linear Equations

✎ **Write the equation of the line through the given points.**

1) Through: $(6, -10), (10, 14)$

2) Through: $(10, 4), (4, 22)$

3) Through: $(-6, 4), (2, 12)$

4) Through: $(15, 11), (3, -1)$

5) Through: $(-5, 33), (9, 5)$

6) Through: $(20, 5), (17, 2)$

7) Through: $(24, -4), (16, 4)$

8) Through: $(-18, 57), (33, -45)$

9) Through: $(10, 12), (8, 18)$

10) Through: $(25, 41), (33, -7)$

11) Through: $(-6, 9), (-8, -7)$

12) Through: $(8, 8), (4, -8)$

13) Through: $(6, -10), (10, 6)$

14) Through: $(10, -24), (-8, 12)$

15) Through: $(10, 10), (-2, -4)$

16) Through: $(-7, 35), (11, -31)$

✎ **Find the answer for each problem.**

17) What is the equation of a line with slope 3 and intercept 11?

_____

18) What is the equation of a line with slope 5 and intercept 15?

_____

19) What is the equation of a line with slope 7 and passes through point $(3, 2)$? _____

20) What is the equation of a line with slope -3 and passes through point $(-2, 5)$? _____

21) The slope of a line is -6 and it passes through point $(-2, 1)$. What is the equation of the line? _____

22) The slope of a line is 5 and it passes through point $(-4, 2)$. What is the equation of the line? _____

Graphing Linear Inequalities

✎ **Sketch the graph of each linear inequality.**

1) $y > 3x - 5$ 2) $y < 2x + 1$ 3) $y \leq -4x - 5$

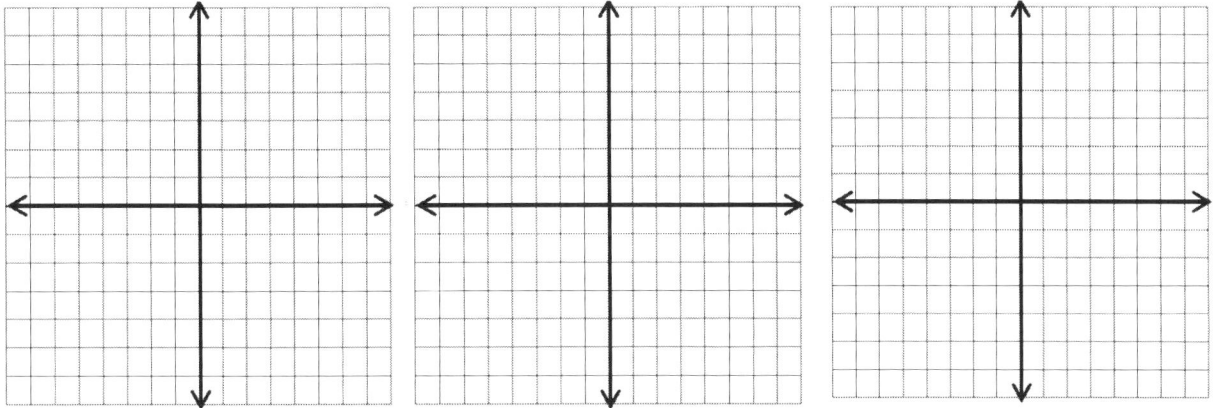

4) $2y \geq 12 + 4x$ 5) $-5y < x - 15$ 6) $3y \geq -9x + 6$

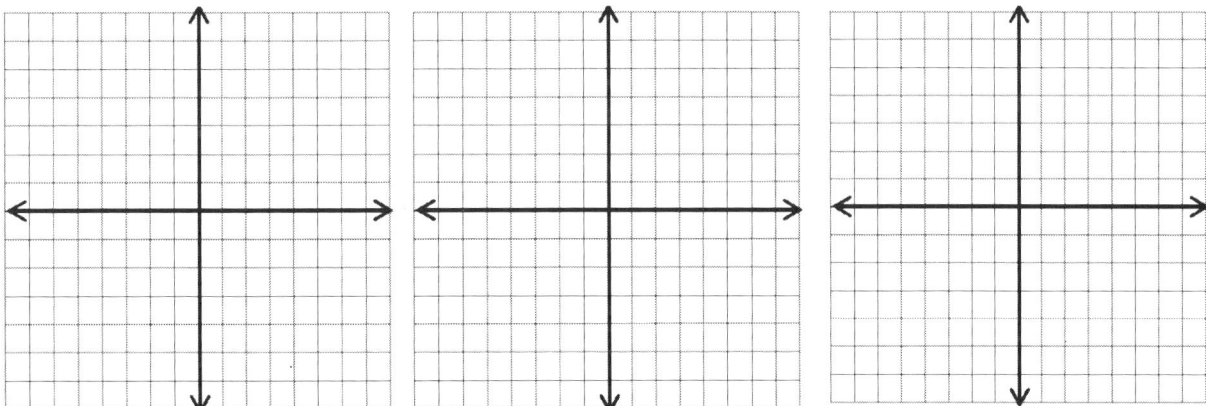

Finding Midpoint

✎ **Find the midpoint of the line segment with the given endpoints.**

1) $(-4, -6), (2, 4)$

2) $(13, 5), (-1, 5)$

3) $(11, -4), (3, 14)$

4) $(-15, -6), (3, 9)$

5) $(7, -8), (13, -12)$

6) $(-14, -8), (8, -12)$

7) $(9, 2), (-9, 22)$

8) $(-8, 10), (-8, 4)$

9) $(-7, 7), (23, -15)$

10) $(3, 17), (19, -5)$

11) $(-4, 13), (7, 9)$

12) $(11, 8), (-3, -6)$

13) $(-4, 12), (0, 6)$

14) $(34, 12), (18, -28)$

15) $(15, 6), (-1, 0)$

16) $(-11, -13), (-13, 19)$

17) $(12, 4), (8, 16)$

18) $(-2, -7), (18, -21)$

19) $(18, 13), (-6, 5)$

20) $(10, -4), (0, 18)$

21) $(4, -4), (8, -20)$

22) $(25, 5), (-11, -17)$

23) $(8, 12), (16, -2)$

24) $(14, -20), (8, 14)$

✎ **Find the answer for each problem.**

25) One endpoint of a line segment is $(6, 8)$ and the midpoint of the line segment is $(1, 6)$. What is the other endpoint? _____

26) One endpoint of a line segment is $(-7, 5)$ and the midpoint of the line segment is $(1, 3)$. What is the other endpoint? _____

27) One endpoint of a line segment is $(-6, -10)$ and the midpoint of the line segment is $(2, 9)$. What is the other endpoint? _____

Finding Distance of Two Points

✎ **Find the distance between each pair of points.**

1) $(5, 9), (-11, -3)$

2) $(-6, 2), (-2, 6)$

3) $(-8, -1), (-3, 8)$

4) $(-8, -2), (2, 22)$

5) $(6, -4), (-12, -28)$

6) $(-6, 0), (-2, 3)$

7) $(8, 12), (8, 6)$

8) $(12, -10), (12, -2)$

9) $(15, 27), (-33, -9)$

10) $(10, -2), (6, -14)$

11) $(1, 0), (6, 12)$

12) $(8, 4), (3, -8)$

13) $(3, 2), (-5, -11)$

14) $(-10, 12), (6, 42)$

15) $(0, 16), (-8, 10)$

16) $(5, 0), (30, 60)$

17) $(3, 5), (-5, -10)$

18) $(-4, 6), (4, 3)$

19) $(7, 2), (-8, -18)$

20) $(-10, 8), (14, 18)$

✎ **Find the answer for each problem.**

21) Triangle ABC is a right triangle on the coordinate system and its vertices are $(-5, 7), (-5, 1),$ and $(1, 1).$ What is the area of triangle ABC?

_____

22) Three vertices of a triangle on a coordinate system are $(1, 1), (7, 1),$ and $(1, 9).$ What is the perimeter of the triangle? _____

23) Four vertices of a rectangle on a coordinate system are $(-2, 4), (-2, 7),$ $(4, 4),$ and $(4, 7).$ What is its perimeter? _____

Answers of Worksheets – Chapter 7

Finding Slope

1) 2

2) -1

3) 4

4) -3

5) 7

6) -4

7) 7

8) -8

9) -9

10) 8

11) $\frac{1}{5}$

12) $-\frac{3}{7}$

13) $\frac{1}{2}$

14) -1

15) $\frac{3}{8}$

16) $\frac{3}{11}$

17) 2

18) 1

19) $-\frac{5}{2}$

20) 3

21) -1

22) 3

23) 2

24) -3

25) -1

26) 3

27) 3

28) -11

Graphing Lines Using Line Equation

1) $y = x - 5$

2) $y = -3x + 4$

3) $x - 2y = 0$

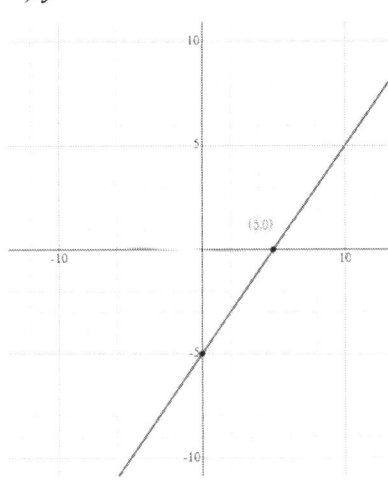

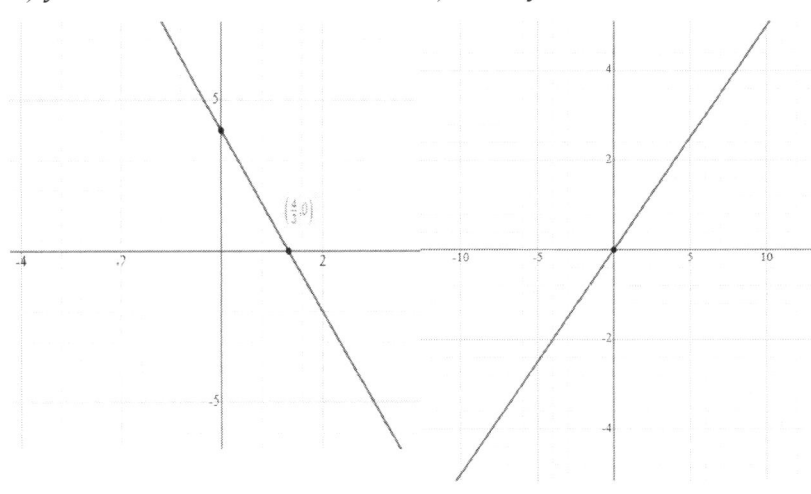

4) $x + y = -4$

5) $4x + 3y = -2$

6) $y - 3x + 6 = 0$

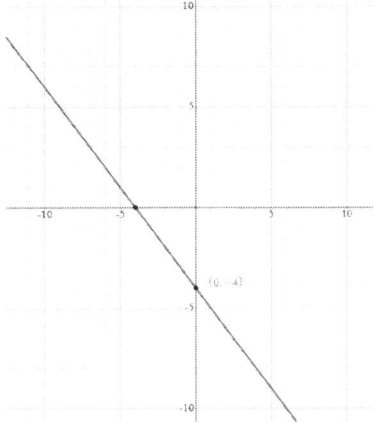

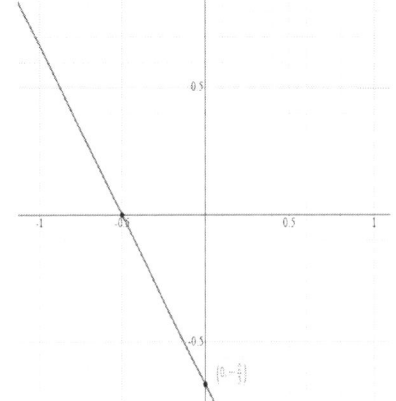

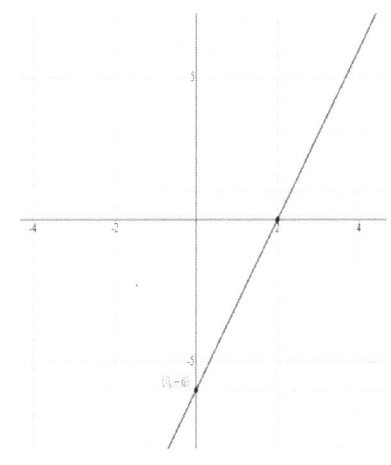

Writing Linear Equations

1) $y = 6x - 46$

2) $y = -3x + 34$

3) $y = x + 10$

4) $y = x - 4$

5) $y = -2x + 23$

6) $y = x - 15$

7) $y = -x + 20$

8) $y = -2x + 21$

9) $y = -3x + 42$

10) $y = -6x + 191$

11) $y = 8x + 57$

12) $y = 4x - 24$

13) $y = 4x - 34$

14) $y = -2x - 4$

15) $y = \frac{7}{6}x - \frac{5}{3}$

16) $y = -\frac{11}{3}x + \frac{28}{3}$

17) $y = 3x + 11$

18) $y = 5x + 15$

19) $y = 7x - 19$

20) $y = -3x - 1$

21) $y = -6x - 11$

22) $y = 5x + 22$

Graphing Linear Inequalities

1) $y > 3x - 5$

2) $y < 2x + 1$

3) $y \leq -4x - 5$

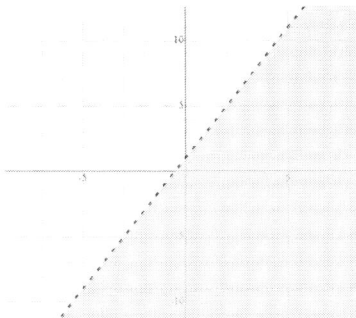
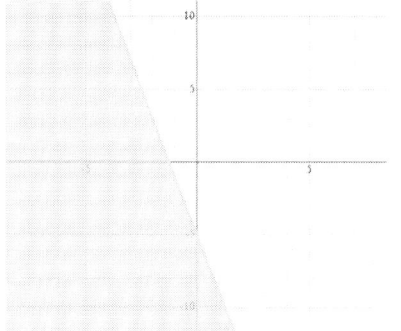
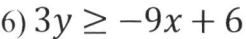

4) $2y \geq 12 + 4x$

5) $-5y < x - 15$

6) $3y \geq -9x + 6$

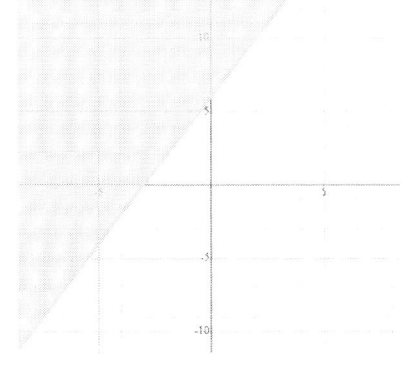
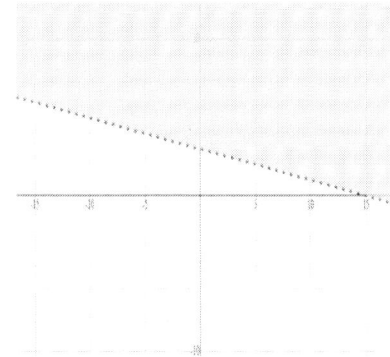
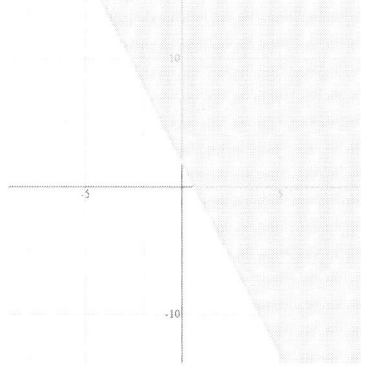

Finding Midpoint

1) $(-1, -1)$

2) $(6, 5)$

3) $(7, 5)$

4) $(-6, 1.5)$

5) $(10, -10)$

6) $(-3, -10)$

7) $(0, 12)$

8) $(-8, 7)$

9) $(8, -4)$

10) $(11, 6)$

11) $(1.5, 11)$

12) $(4, 1)$

13) $(-2, 9)$

14) $(26, -8)$

15) $(7, 3)$

16) $(-12, 3)$

17) $(10, 10)$

18) $(8, -14)$

19) $(6, 9)$

20) $(5, 7)$

21) $(6, -12)$

22) $(7, -6)$

23) $(12, 5)$

24) $(11, -3)$

25) $(-4, 4)$

26) $(9, 1)$

27) $(10, 28)$

Finding Distance of Two Points

1) 20

2) $4\sqrt{2}$

3) $\sqrt{106}$

4) 26

5) 30

6) 10

7) 6

8) 8

9) 60

10) $4\sqrt{10}$

11) 13

12) 13

13) $\sqrt{233}$

14) 34

15) 10

16) 65

17) 17

18) $\sqrt{73}$

19) 25

20) 26

21) 18 square units

22) 24 units

23) 18 units

Chapter 8:

Polynomials

Topics that you will practice in this chapter:

- ✓ Writing Polynomials in Standard Form
- ✓ Simplifying Polynomials
- ✓ Adding and Subtracting Polynomials
- ✓ Multiplying Monomials
- ✓ Multiplying and Dividing Monomials
- ✓ Multiplying a Polynomial and a Monomial
- ✓ Multiplying Binomials
- ✓ Factoring Trinomials
- ✓ Operations with Polynomials

Mathematics is the supreme judge; from its decisions there is no appeal. – Tobias Dantzig

Writing Polynomials in Standard Form

✎ **Write each polynomial in standard form.**

1) $9x - 7x =$

2) $-6 + 15x - 15x =$

3) $3x^2 - 11x^3 =$

4) $18 + 19x^3 - 14 =$

5) $3x^2 + 9x - 4x^5 =$

6) $-7x^3 + 12x^7 =$

7) $9x + 6x^2 - 2x^6 =$

8) $-5x^3 + x - 9x^4 =$

9) $8x^2 + 34 - 21x =$

10) $8 - 7x + 11x^4 =$

11) $25x^3 + 45x - 13x^4 =$

12) $17 + 9x^2 - 2x^3 =$

13) $18x^2 - 8x + 8x^3 =$

14) $9x^4 - 4x^2 - 10x^5 =$

15) $-41 + 7x^2 - 8x^4 =$

16) $8x^2 - 7x^5 + 3x^3 - 12 =$

17) $4x^2 - 9x^5 + 12 - 8x^4 =$

18) $-2x^5 + 6x - 9x^2 - 7x =$

19) $14x^5 + 7x^4 - 8x^5 - 8x^2 =$

20) $2x^3 - 15x^4 + 9x^3 + 3x^8 =$

21) $7x^4 - 16x^5 - 9x^2 + 10x^4 =$

22) $5x^2 + 6x^5 + 37x^3 - 9x^5 =$

23) $3x(2x + 5 - 6x^2) =$

24) $12x(x^6 + 2x^3) =$

25) $6x(x^2 + 8x + 4) =$

26) $8x(3 - 2x + 4x^3) =$

27) $7x(2x^3 - 2x^2 + 2) =$

28) $5x(5x^5 + 4x^4 - 1) =$

29) $x(4x^3 + 52x^4 + 2x) =$

30) $6x(3x - 4x^4 + 7x^2) =$

Simplifying Polynomials

✍ **Simplify each expression.**

1) $3(x - 12) =$

2) $5x(2x - 4) =$

3) $7x(5x - 1) =$

4) $6x(3x + 2) =$

5) $5x(2x - 7) =$

6) $9x(x + 8) =$

7) $(3x - 8)(x - 3) =$

8) $(x - 9)(3x + 4) =$

9) $(x - 8)(x - 5) =$

10) $(3x + 4)(3x - 4) =$

11) $(5x - 8)(5x - 2) =$

12) $7x^2 + 7x^2 - 6x^4 =$

13) $5x - 2x^2 + 7x^3 + 10 =$

14) $8x + 2x^2 - 5x^3 =$

15) $15x + 4x^5 - 8x^2 =$

16) $-4x^2 + 7x^5 + 11x^4 =$

17) $-14x^2 + 8x^3 - 2x^4 + 5x =$

18) $14 - 5x^2 + 6x^2 - 10x^3 + 17 =$

19) $x^2 - 9x + 2x^3 + 15x - 10x =$

20) $14 - 8x^2 + 4x^2 - 9x^3 + 1 =$

21) $-4x^5 + 2x^4 - 18x^2 + 2x^5 =$

22) $(3x^3 - 5) + (3x^3 - 2x^3) =$

23) $4(3x^5 - 3x^3 - 6x^5) =$

24) $-4(x^5 + 8) - 4(12 - x^5) =$

25) $7x^2 - 9x^3 - 2x + 14 - 5x^2 =$

26) $10 - 5x^2 + 3x^2 - 4x^3 + 4 =$

27) $(8x^2 - 2x) - (5x - 5 - 4x^2) =$

28) $4x^4 - 8x^3 - x(3x^2 + 5x) =$

29) $4x + 8x^2 - 10 - 2(x^2 - 1) =$

30) $5 - 3x^2 + (6x^4 - 2x^2 + 8x^4) =$

31) $-(x^5 + 8) - 7(4 + x^5) =$

32) $(4x^3 - x) - (x - 6x^3) =$

Adding and Subtracting Polynomials

✎ **Add or subtract expressions.**

1) $(-x^3 - 3) + (4x^3 + 2) =$

2) $(3x^2 + 4) - (6 - x^2) =$

3) $(x^3 + 4x^2) - (5x^3 + 15) =$

4) $(3x^3 - 2x^2) + (2x^2 - x) =$

5) $(10x^3 + 14x) - (14x^3 + 7) =$

6) $(5x^2 - 7) + (3x^2 + 7) =$

7) $(9x^3 + 4) - (10 - 5x^3) =$

8) $(x^2 + 2x^3) - (2x^3 + 5) =$

9) $(8x^2 - x) + (5x - 4x^2) =$

10) $(17x + 10) - (2x + 10) =$

11) $(12x^4 - 4x) - (x - 3x^4) =$

12) $(3x - x^4) - (7x^4 + 8x) =$

13) $(7x^3 - 6x^5) - (4x^5 - 2x) =$

14) $(x^3 - 7) + (4x^3 + 8x^5) =$

15) $(6x^2 + 5x^4) - (x^4 - 9x^2) =$

16) $(-4x^2 - 4x) + (7x - 8x^2) =$

17) $(x - 6x^4) - (15x^4 + 2x) =$

18) $(4x - 3x^4) - (2x^4 - 3x^3) =$

19) $(7x^3 - 7) + (6x^3 - 6x^2) =$

20) $(9x^5 + 7x^4) - (x^4 - 5x^5) =$

21) $(-4x^2 + 11x^4 + 2x^3) + (20x^3 + 4x^4 + 12x^2) =$

22) $(5x^2 - 5x^4 - 5x) - (-4x^2 - 5x^4 + 5x) =$

23) $(12x + 36x^3 - 10x^4) + (20x^3 + 10x^4 - 7x) =$

24) $(2x^5 - 4x^3 - 5x) - (2x^2 + 7x^3 - 2x) =$

25) $(14x^3 - 4x^5 - x) - (-4x^3 - 12x^5 + 9x) =$

26) $(-5x^2 + 12x^4 + x^3) + (10x^3 + 17x^4 + 7x^2) =$

Multiplying Monomials

✎ **Simplify each expression.**

1) $7u^5 \times (-u^2) =$

2) $(-9p^8) \times (-4p^2) =$

3) $5xy^3z^3 \times 4z^2 =$

4) $8u^6t \times 2ut^2 =$

5) $(-2a^2) \times (-5a^3b^3) =$

6) $-4a^2b^2 \times 5a^4b =$

7) $10xy^4 \times 2x^2y^2 =$

8) $4p^2q^4 \times (-2pq^2) =$

9) $8s^5t^4 \times 3st^4 =$

10) $(-7x^5y^3) \times 7x^4y =$

11) $xy^7z \times 15z^3 =$

12) $15xy \times 2x^3y =$

13) $14pq^4 \times (-3p^3q) =$

14) $25s^4t^2 \times st^6 =$

15) $12p^5 \times (-2p^3) =$

16) $(-12p^2q^4r) \times 3pq^5r^3 =$

17) $(-7a^4) \times (-4a^5b) =$

18) $4u^7v^2 \times (-9u^4v^6) =$

19) $9u^5 \times (-3u) =$

20) $-3xy^9 \times 8x^5y =$

21) $12y^5z^3 \times (-2 y^2z) =$

22) $9a^3bc^5 \times 4abc^3 =$

23) $(-9p^5q^2) \times (-3p^2q^4) =$

24) $4u^8v^3 \times (-4u^8v^5) =$

25) $15y^3z^4 \times (-y^5z) =$

26) $(-12pq^4r^3) \times 5p^4q^2r =$

27) $3ab^5c^2 \times 3a^2bc^4 =$

28) $7x^5yz^3 \times 9x^5y^7z^4 =$

Multiplying and Dividing Monomials

✍ **Simplify each expression.**

1) $(7x^2)(x^3) =$

2) $(4x^3)(5x^2) =$

3) $(3x^4)(2x^2) =$

4) $(5x^8)(8x^3) =$

5) $(12x^6)(2x^3) =$

6) $(2yx^5)(16x^2) =$

7) $(9x^5y)(2x^2y^3) =$

8) $(-2x^2y^5)(5x^3y^2) =$

9) $(-4x^2y^2)(-8x^4y^3) =$

10) $(2x^4y)(-5x^5y^3) =$

11) $(9x^4y^4)(2x^3y^3) =$

12) $(2x^4y^6)(3x^4y^3) =$

13) $(8x^3y^8)(7x^5y^{10}) =$

14) $(14x^6y^5)(3x^5y^5) =$

15) $(8x^2y^8)(5x^{10}y^{10}) =$

16) $(-3x^2y^5)(4x^6y^3) =$

17) $\dfrac{9x^4y^5}{xy^3} =$

18) $\dfrac{18x^8y^3}{18x^7y} =$

19) $\dfrac{54x^4y^4}{6xy} =$

20) $\dfrac{63x^4y^5}{7x^3y^4} =$

21) $\dfrac{32x^7y^6}{8x^2y^3} =$

22) $\dfrac{63x^9y^4}{3x^4y^3} =$

23) $\dfrac{96x^{16}y^{12}}{12x^7y^9} =$

24) $\dfrac{60x^{10}y^6}{12x^{11}y^3} =$

25) $\dfrac{90x^8y^{12}}{18x^7y^{12}} =$

26) $\dfrac{45x^{23}y^{10}}{9x^9y^6} =$

27) $\dfrac{-96x^8y^8}{24x^6y^8} =$

Multiplying a Polynomial and a Monomial

✎ **Find each product.**

1) $2x(x + 4) =$

2) $3(8 - x) =$

3) $5x(3x + 4) =$

4) $x(-2x + 5) =$

5) $7x(3x - 3) =$

6) $3(2x - 5y) =$

7) $6x(7x - 3) =$

8) $x(12x + 5y) =$

9) $5x(x + 6y) =$

10) $11x(4x + 5y) =$

11) $8x(4x + 2) =$

12) $12x(x - 15y) =$

13) $9x(5x - 3y) =$

14) $8x(5x - 2y + 5) =$

15) $9x(2x^2 + 7y^2) =$

16) $8x(9x + 6y) =$

17) $2(3x^5 - 2y^5) =$

18) $4x(-x^2y + 2y) =$

19) $-3(2x^3 - 3xy + 9) =$

20) $2(x^2 - 2xy - 4) =$

21) $7x(4x^3 - xy + 2x) =$

22) $-9x(-2x^3 - 2x + 7xy) =$

23) $6(x^2 + 3xy - 8y^2) =$

24) $5x(7x^3 - x + 8) =$

25) $7(x^{24} - 4x - 6) =$

26) $x^2(-3x^3 + 4x + 7) =$

27) $x^2(2x^3 + 10 - 5x) =$

28) $4x^4(3x^3 - 2x + 8) =$

29) $5x^2(x^4 - 5xy + 2y^3) =$

30) $4x^2(7x^4 - 2x + 11) =$

31) $7x^3(3x^3 + 5x - 7) =$

32) $4x(x^2 - 8xy + 7y^3) =$

Multiplying Binomials

✎ **Find each product.**

1) $(x + 5)(x + 1) =$

2) $(x - 3)(x + 7) =$

3) $(x - 1)(x - 9) =$

4) $(x + 3)(x + 8) =$

5) $(x - 4)(x - 11) =$

6) $(x + 5)(x + 6) =$

7) $(x - 8)(x + 7) =$

8) $(x - 3)(x - 2) =$

9) $(x + 8)(x + 11) =$

10) $(x - 3)(x + 5) =$

11) $(x + 8)(x + 8) =$

12) $(x + 2)(x + 7) =$

13) $(x - 9)(x + 4) =$

14) $(x - 10)(x + 10) =$

15) $(x + 24)(x + 2) =$

16) $(x + 9)(x + 13) =$

17) $(x - 7)(x + 7) =$

18) $(x - 5)(x + 2) =$

19) $(3x + 4)(x + 5) =$

20) $(x - 8)(5x + 2) =$

21) $(x - 9)(4x + 9) =$

22) $(2x - 7)(3x - 2) =$

23) $(x - 4)(x + 11) =$

24) $(5x - 6)(2x + 4) =$

25) $(4x - 9)(x + 7) =$

26) $(8x - 5)(2x + 2) =$

27) $(3x + 9)(7x + 4) =$

28) $(6x - 8)(4x + 4) =$

29) $(4x + 5)(5x - 8) =$

30) $(8x - 1)(8x + 4) =$

31) $(9x + 4)(3x - 6) =$

32) $(4x^2 + 12)(4x^2 - 12) =$

Factoring Trinomials

✎ **Factor each trinomial.**

1) $x^2 + 12x + 35 =$

2) $x^2 - 8x + 12 =$

3) $x^2 + 11x + 10 =$

4) $x^2 - 12x + 27 =$

5) $x^2 - 16x + 15 =$

6) $x^2 - 13x + 40 =$

7) $x^2 + 15x + 44 =$

8) $x^2 + x - 72 =$

9) $x^2 - 81 =$

10) $x^2 - 17x + 70 =$

11) $x^2 + 8x - 48 =$

12) $x^2 + 5x - 104 =$

13) $x^2 - 7x - 18 =$

14) $x^2 + 22x + 121 =$

15) $3x^2 - 3x - 36 =$

16) $2x^2 - 35x + 75 =$

17) $14x^2 + 11x - 15 =$

18) $8x^2 - 12x - 20 =$

19) $15x^2 + 16x + 4 =$

20) $24x^2 + 2x - 1 =$

✎ **Calculate each problem.**

21) The area of a rectangle is $x^2 - 3x - 40$. If the width of rectangle is $x - 8$, what is its length? _____

22) The area of a parallelogram is $12x^2 + 7x - 10$ and its height is $4x + 5$. What is the base of the parallelogram? _____

23) The area of a rectangle is $10x^2 - 43x + 28$. If the width of the rectangle is $5x - 4$, what is its length? _____

Operations with Polynomials

✏ **Find each product.**

1) $2(4x + 1) = $ _____

2) $5(2x + 7) = $ _____

3) $4(6x - 5) = $ _____

4) $-4(7x - 8) = $ _____

5) $3x^2(8x + 4) = $ _____

6) $6x^2(2x - 9) = $ _____

7) $5x^3(-x + 4) = $ _____

8) $-5x^4(4x - 9) = $ _____

9) $6(x^2 + 7x - 3) = $ _____

10) $4(3x^2 - 2x + 6) = $ _____

11) $9(3x^2 + 8x + 2) = $ _____

12) $7x(x^2 + 5x + 3) = $ _____

13) $(7x + 2)(2x - 5) = $ _____

14) $(8x + 5)(3x - 8) = $ _____

15) $(4x + 2)(6x - 1) = $ _____

16) $(5x - 4)(5x + 9) = $ _____

✏ **Calculate each problem.**

17) The measures of two sides of a triangle are $(2x + 8y)$ and $(5x - 3y)$. If the perimeter of the triangle is $(11x + 6y)$, what is the measure of the third side? _____

18) The height of a triangle is $(8x + 2)$ and its base is $(2x - 6)$. What is the area of the triangle? _____

19) One side of a square is $(4x + 3)$. What is the area of the square? _____

20) The length of a rectangle is $(7x - 9y)$ and its width is $(13x + 9y)$. What is the perimeter of the rectangle? _____

21) The side of a cube measures $(x + 2)$. What is the volume of the cube? _____

22) If the perimeter of a rectangle is $(24x + 10y)$ and its width is $(4x + 3y)$, what is the length of the rectangle? _____

Answers of Worksheets – Chapter 8

Writing Polynomials in Standard Form

1) $2x$

2) -6

3) $-11x^3 + 3x^2$

4) $19x^4 + 4$

5) $-4x^5 + 3x^2 + 9x$

6) $12x^7 - 7x^3$

7) $-2x^6 + 6x^2 + 9x$

8) $-9x^4 - 5x^3 + x$

9) $8x^2 - 21x + 34$

10) $11x^4 - 7x + 8$

11) $-13x^4 + 25x^3 + 45x$

12) $-2x^3 + 9x^2 + 17$

13) $8x^3 + 18x^2 - 8x$

14) $-10x^5 - 9x^4 - 4x^2$

15) $-8x^4 + 7x^2 - 41$

16) $-7x^5 + 3x^3 + 8x^2 - 12$

17) $-9x^5 - 8x^4 + 4x^2 + 12$

18) $-2x^5 - 9x^2 - x$

19) $6x^5 + 7x^4 - 8x^2$

20) $3x^8 - 15x^4 + 11x^2$

21) $-16x^5 + 17x^4 - 9x^2$

22) $-3x^5 + 37x^3 + 5x^2$

23) $-18x^3 + 6x^2 + 15x$

24) $12x^7 + 24x^4$

25) $6x^3 + 48x^2 + 24x$

26) $32x^4 - 16x^2 + 24x$

27) $14x^4 - 14x^3 + 14x$

28) $25x^6 + 20x^5 - 5x$

29) $52x^5 + 4x^4 + 2x^2$

30) $-24x^5 + 42x^3 + 18x^2$

Simplifying Polynomials

1) $3x - 36$

2) $10x^2 - 20x$

3) $35x^2 - 7x$

4) $18x^2 + 12x$

5) $10x^2 - 35x$

6) $9x^2 + 72x$

7) $3x^2 - 17x + 24$

8) $3x^2 - 23x - 36$

9) $x^2 - 13x + 40$

10) $9x^2 - 16$

11) $25x^2 - 50x + 16$

12) $-6x^4 + 14x^2$

13) $7x^3 - 2x^2 + 5x + 10$

14) $-5x^3 + 2x^2 + 8x$

15) $4x^5 - 8x^2 + 15x$

16) $7x^5 + 11x^4 - 4x^2$

17) $-2x^4 + 8x^3 - 14x^2 + 5x$

18) $-10x^3 + x^2 + 31$

19) $2x^3 + x^2 - 4x$

20) $-9x^3 - 4x^2 + 15$

21) $-2x^5 + 2x^4 - 18x^2$

22) $4x^3 - 5$

23) $-12x^5 - 12x^3$

24) -80

25) $-9x^3 + 2x^2 - 2x + 14$

26) $-4x^3 - 2x^2 + 14$

27) $12x^2 - 7x + 5$

28) $4x^4 - 11x^3 - 5x^2$

29) $6x^2 + 4x - 8$

30) $14x^4 - 5x^2 + 5$

31) $-8x^5 - 36$

32) $10x^3 - 2x$

Adding and Subtracting Polynomials

1) $3x^2 - 1$

2) $4x^2 - 2$

3) $-4x^3 + 4x^2 - 15$

4) $3x^3 - x$

5) $-4x^3 + 14x - 7$

6) $8x^2$

7) $14x^3 - 6$

8) $x^2 - 5$

9) $4x^2 + 4x$

10) $15x$

11) $15x^4 - 5x$

12) $-8x^4 - 5x$

13) $-10x^5 + 7x^3 + 2x$

14) $5x^5 + 5x^3 - 7$

15) $4x^4 + 15x^2$

16) $-12x^2 + 3x$

17) $-21x^4 - x$

18) $-5x^4 + 3x^3 + 4x$

19) $13x^3 - 6x^2 - 7$

20) $14x^5 + 6x^4$

21) $15x^4 + 22x^3 + 8x^2$

22) $9x^2 - 10x$

23) $56x^3 + 5x$

24) $2x^5 - 11x^3 - 2x^2 - 3x$

25) $8x^5 + 18x^3 - 10x$

26) $29x^4 + 11x^3 + 2x^2$

Multiplying Monomials

1) $-7u^7$

2) $36p^{10}$

3) $20xy^3z^5$

4) $16u^7t^3$

5) $10a^5b^3$

6) $-20a^6b^3$

7) $20x^3y^6$

8) $-8p^3q^6$

9) $24s^6t^8$

10) $-49x^9y^4$

11) $15xy^7z^4$

12) $30x^4y^2$

13) $-42p^4q^5$

14) $25s^5t^8$

15) $-24p^8$

16) $-36p^3q^9r^4$

17) $28a^9b$

18) $-36u^{11}v^8$

19) $-27u^6$

20) $-24x^6y^{10}$

21) $-24y^7z^4$

22) $36a^4b^2c^8$

23) $27p^7q^6$

24) $-16u^{16}v^8$

25) $-15y^8z^5$

26) $-60p^5q^6r^4$

27) $9a^3b^6c^6$

28) $63x^{10}y^8z^7$

Multiplying and Dividing Monomials

1) $7x^5$

2) $20x^5$

3) $6x^6$

4) $40x^{11}$

5) $24x^9$

6) $32x^7y$

7) $18x^7y^4$

8) $-10x^5y^7$

9) $32x^6y^5$

10) $-10x^9y^4$

11) $18x^7y^7$

12) $6x^8y^9$

13) $56x^8y^{18}$

14) $42x^{11}y^{10}$

15) $40x^{12}y^{18}$

16) $-12x^8y^8$

17) $9x^3y^2$

18) xy^2

19) $9x^3y^3$

20) $9xy$

21) $4x^5y^3$

22) $21x^5y$

23) $8x^9y^3$

24) $5x^{-1}y^3$

25) $5x$

26) $5x^{14}y^4$

27) $-4x^2$

Multiplying a Polynomial and a Monomial

1) $2x^2 + 8x$

2) $-3x + 24$

3) $15x^2 + 20x$

4) $-2x^2 + 5x$

5) $21x^2 - 21x$

6) $6x - 15y$

7) $42x^2 - 18x$

8) $12x^2 + 5xy$

9) $5x^2 + 30xy$

10) $44x^2 + 55xy$

11) $32x^2 + 16x$

12) $12 - 180xy$

13) $45x^2 - 27xy$

14) $40x^2 - 16xy + 40x$

15) $18x^3 + 63xy^2$

16) $72x^2 + 48xy$

17) $6x^5 - 2y^5$

18) $-4x^3y + 8xy$

19) $-6x^3 + 9xy - 27$

20) $2x^2 - 4xy - 8$

21) $28x^4 - 7x^2y + 14x^2$

22) $18x^4 + 18x^2 - 63x^2y$

23) $6x^2 + 18xy - 48y^2$

24) $35x^4 - 5x^2 + 40x$

25) $7x^{24} - 28x - 42$

26) $-3x^5 + 4x^3 + 7x^2$

27) $2x^5 - 5x^3 + 10x^2$

28) $12x^7 - 8x^5 + 32x^4$

29) $5x^6 - 25x^3y + 10x^2y^3$

30) $28x^6 - 8x^3 + 44x^2$

31) $21x^6 + 35x^4 - 49x^3$

32) $4x^3 - 32x^2y + 28xy^3$

Multiplying Binomials

1) $x^2 + 6x + 5$

2) $x^2 + 4x - 21$

3) $x^2 - 10x + 9$

4) $x^2 + 11x + 24$

5) $x^2 - 15x + 44$

6) $x^2 + 11x + 30$

7) $x^2 - x - 56$

8) $x^2 - 5x + 6$

9) $x^2 + 19x + 88$

10) $x^2 + 2x + 15$

11) $x^2 + 16x + 64$

12) $x^2 + 9x + 14$

13) $x^2 - 5x - 36$

14) $x^2 - 100$

15) $x^2 + 26x + 48$

16) $x^2 + 22x + 117$

17) $x^2 - 49$

18) $x^2 - 3x - 10$

19) $3x^2 + 19x + 20$

20) $5x^2 - 38x - 16$

21) $4x^2 - 27x - 81$

22) $6x^2 - 25x + 14$

23) $x^2 + 7x - 44$

24) $10x^2 + 8x - 24$

25) $4x^2 + 19x - 63$

26) $16x^2 + 6x - 10$

27) $21x^2 + 75x + 36$

28) $24x^2 - 8x - 32$

29) $20x^2 - 7x - 40$

30) $64x^2 + 24x - 4$

31) $27x^2 - 42x - 24$

32) $16x^4 - 144$

Factoring Trinomials

1) $(x + 5)(x + 7)$

2) $(x - 2)(x - 6)$

3) $(x + 1)(x + 10)$

4) $(x - 9)(x - 3)$

5) $(x - 1)(x - 15)$

6) $(x - 5)(x - 8)$

7) $(x + 4)(x + 11)$

8) $(x + 9)(x - 8)$

9) $(x - 9)(x + 9)$

10) $(x - 7)(x - 10)$

11) $(x - 4)(x + 12)$

12) $(x - 8)(x + 13)$

13) $(x + 2)(x - 9)$

14) $(x + 11)(x + 11)$

15) $(3x + 9)(x - 4)$

16) $(x - 15)(2x - 5)$

17) $(7x - 5)(2x + 3)$

18) $(2x - 5)(4x + 4)$

19) $(3x + 2)(5x + 2)$

20) $(6x - 1)(4x + 1)$

21) $(x + 5)$

22) $(3x - 2)$

23) $(2x - 7)$

Operations with Polynomials

1) $8x + 2$

2) $10x + 35$

3) $24x - 20$

4) $-28x + 32$

5) $24x^3 + 12x^2$

6) $12x^3 - 54x^2$

7) $-5x^4 + 20x^3$

8) $-20x^5 + 45x^4$

9) $6x^2 + 42x - 18$

10) $12x^2 - 8x + 24$

11) $27x^2 + 72x + 18$

12) $7x^3 + 35x^2 + 21x$

13) $14x^2 - 31x - 10$

14) $24x^2 - 49x - 40$

15) $24x^2 + 8x - 2$

16) $25x^2 + 25x - 36$

17) $(4x + y)$

18) $8x^2 - 22x - 6$

19) $16x^2 + 24x + 9$

20) $40x$

21) $x^3 + 6x^2 + 12x + 8$

22) $(8x + 2y)$

Chapter 9: Functions Operations and Quadratic

Topics that you will practice in this chapter:

- ✓ Graphing Quadratic Functions
- ✓ Solving Quadratic Equations
- ✓ Use the Quadratic Formula and the Discriminant
- ✓ Solve Quadratic Inequalities
- ✓ Evaluating Function
- ✓ Adding and Subtracting Functions
- ✓ Multiplying and Dividing Functions
- ✓ Composition of Functions

It's fine to work on any problem, so long as it generates interesting mathematics along the way – even if you don't solve it at the end of the day." – Andrew Wiles

Evaluating Function

✍ Write each of following in function notation.

1) $h = -8x + 9$

2) $k = 5a - 21$

3) $d = 14t$

4) $y = \frac{3}{17}x - \frac{9}{17}$

5) $m = 18n - 94$

6) $c = p^2 - 7p + 15$

✍ Evaluate each function.

7) $f(x) = 6x - 7$, find $f(-3)$

8) $g(x) = \frac{1}{10}x + 6$, find $f(5)$

9) $h(x) = -2x + 15$, find $f(8)$

10) $f(x) = -3x + 8$, find $f(-2)$

11) $f(a) = 12a - 9$, find $f(0)$

12) $h(x) = 18 - 5x$, find $f(-4)$

13) $g(n) = 7n - 5$, find $f(5)$

14) $f(x) = -9x - 2$, find $f(3)$

15) $k(n) = -12 + 4.5n$, find $f(2)$

16) $f(x) = -1.5x + 2.5$, find $f(-6)$

17) $g(n) = \frac{16n-8}{6n}$, find $g(2)$

18) $g(n) = \sqrt{5n} - 2$, find $g(5)$

19) $h(x) = x^{-1} - 6$, find $h(\frac{1}{9})$

20) $h(n) = n^{-3} + 4$, find $h(\frac{1}{2})$

21) $h(n) = n^2 - 5$, find $h(\frac{4}{5})$

22) $h(n) = n^3 - 8$, find $h(-\frac{1}{3})$

23) $h(n) = 4n^2 - 42$, find $h(-4)$

24) $h(n) = -5n^2 - 9n$, find $h(7)$

25) $g(n) = \sqrt{4n^2} - \sqrt{5n}$, find $g(5)$

26) $h(a) = \frac{-15a+7}{3a}$, find $h(-b)$

27) $k(a) = 8a - 9$, find $k(a - 3)$

28) $h(x) = \frac{1}{6}x + 7$, find $h(-12x)$

29) $h(x) = 8x^2 + 10$, find $h(\frac{x}{2})$

30) $h(x) = x^4 - 8$, find $h(-2x)$

Adding and Subtracting Functions

✎ **Perform the indicated operation.**

1) $f(x) = 2x + 3$

   $g(x) = x + 4$

   Find $(f - g)(2)$

2) $g(a) = -3a - 8$

   $f(a) = -4a - 12$

   Find $(g - f)(-2)$

3) $h(t) = 7t + 5$

   $g(t) = 3t + 11$

   Find $(h - g)(t)$

4) $g(a) = -5a - 3$

   $f(a) = 3a^2 + 4$

   Find $(g - f)(x)$

5) $g(x) = \frac{2}{7}x - 10$

   $h(x) = \frac{5}{7}x + 10$

   Find $g(14) - h(14)$

6) $h(3) = \sqrt{7x} - 2$

   $g(x) = \sqrt{7x} + 2$

   Find $(h + g)(7)$

7) $f(x) = x^{-3}$

   $g(x) = x^2 + \frac{4}{x}$

   Find $(f - g)(-1)$

8) $h(n) = n^2 + 8$

   $g(n) = -n + 5$

   Find $(h - g)(a)$

9) $g(x) = -2x^2 - 3 - x$

   $f(x) = 7 + x$

   Find $(g - f)(2x)$

10) $g(t) = 4t - 9$

    $f(t) = -t^2 + 5$

    Find $(g + f)(-z)$

11) $f(x) = 3x + 9$

    $g(x) = -4x^2 + 2x$

    Find $(f - g)(-x^2)$

12) $f(x) = -9x^3 - 4x$

    $g(x) = 4x + 12$

    Find $(f + g)(3x^2)$

Multiplying and Dividing Functions

✎ **Perform the indicated operation.**

1) $g(x) = -2x - 5$

$f(x) = 3x + 4$

Find $(g.f)(2)$

2) $f(x) = 3x$

$h(x) = -2x + 5$

Find $(f.h)(-3)$

3) $g(a) = 5a - 3$

$h(a) = a - 7$

Find $(g.h)(-3)$

4) $f(x) = x - 4$

$h(x) = 4x - 3$

Find $(\frac{f}{h})(4)$

5) $f(x) = 9a^2$

$g(x) = 5 + 4a$

Find $(\frac{f}{g})(3)$

6) $g(a) = \sqrt{5a} + 7$

$f(a) = (-a)^2 + 3$

Find $(\frac{g}{f})(5)$

7) $g(t) = t^2 + 5$

$h(t) = 2t - 5$

Find $(g.h)(-3)$

8) $g(n) = n^2 + 2n - 4$

$h(n) = -n + 6$

Find $(g.h)(1)$

9) $g(a) = (a - 7)^3$

$f(a) = a^2 + 8$

Find $(\frac{g}{f})(7)$

10) $g(x) = -x^2 + \frac{4}{5}x + 10$

$f(x) = x^2 - 3$

Find $(\frac{g}{f})(5)$

11) $f(x) = x^3 - 3x^2 + 9$

$g(x) = x - 4$

Find $(f.g)(x)$

12) $f(x) = 3x - 5$

$g(x) = x^2 - 4x$

Find $(f.g)(x^2)$

Composition of Functions

✎ Using $f(x) = 2x - 5$ and $g(x) = -2x$, find:

1) $f\big(g(0)\big) =$

4) $g\big(f(3)\big) =$

2) $f\big(g(-1)\big) =$

5) $f\big(g(-2)\big) =$

3) $g\big(f(1)\big) =$

6) $g\big(f(5)\big) =$

✎ Using $f(x) = -\frac{1}{4}x + \frac{3}{4}$ and $g(x) = x^2$, find:

7) $g\big(f(4)\big) =$

10) $f\big(f(1)\big) =$

8) $g\big(f(3)\big) =$

11) $g\big(f(-1)\big) =$

9) $g\big(g(2)\big) =$

12) $g\big(f(7)\big) =$

✎ Using $f(x) = -5x + 2$ and $g(x) = x + 3$, find:

13) $g\big(f(0)\big) =$

16) $f\big(g(-3)\big) =$

14) $f\big(f(2)\big) =$

17) $g\big(f(-5)\big) =$

15) $f\big(g(3)\big) =$

18) $f\big(f(x)\big) =$

✎ Using $f(x) = \sqrt{x + 9}$ and $g(x) = x - 9$, find:

19) $f\big(g(9)\big) =$

22) $f\big(f(-5)\big) =$

20) $g\big(f(-8)\big) =$

23) $g\big(f(7)\big) =$

21) $f\big(g(18)\big) =$

24) $g\big(g(8)\big) =$

Quadratic Equation

✎ Multiply.

1) $(x - 2)(x + 8) =$ _____

2) $(x + 1)(x + 9) =$ _____

3) $(x - 5)(x + 6) =$ _____

4) $(x + 7)(x - 3) =$ _____

5) $(x - 9)(x - 8) =$ _____

6) $(4x + 2)(x - 4) =$ _____

7) $(3x - 6)(x + 4) =$ _____

8) $(x - 9)(2x + 7) =$ _____

9) $(5x + 3)(x - 4) =$ _____

10) $(4x + 2)(3x - 3) =$ _____

✎ Factor each expression.

11) $x^2 - 4x - 21 =$ _____

12) $x^2 + 14x + 45 =$ _____

13) $x^2 - 5x - 24 =$ _____

14) $x^2 - 7x + 6 =$ _____

15) $x^2 + 14x + 33 =$ _____

16) $4x^2 + 38x + 18 =$ _____

17) $5x^2 + 18x - 8 =$ _____

18) $2x^2 + 2x - 40 =$ _____

19) $2x^2 + 22x + 56 =$ _____

20) $12x^2 - 148x + 360 =$ _____

✎ Calculate each equation.

21) $(x + 6)(x - 9) = 0$

22) $(x + 1)(x + 11) = 0$

23) $(3x + 9)(x + 3) = 0$

24) $(5x - 5)(6x + 12) = 0$

25) $x^2 - 12x + 30 = 6$

26) $x^2 + 6x + 14 = 5$

27) $x^2 + \frac{9}{2}x + 7 = 5$

28) $x^2 + 2x - 25 = 10$

29) $2x^2 + 12x - 54 = 0$

30) $x^2 - 11x = 12$

Solving Quadratic Equations

✎ **Solve each equation by factoring or using the quadratic formula.**

1) $(x + 5)(x - 2) = 0$

2) $(x + 8)(x + 2) = 0$

3) $(x - 9)(x + 5) = 0$

4) $(x - 3)(x - 1) = 0$

5) $(x + 9)(x + 4) = 0$

6) $(2x + 5)(x + 9) = 0$

7) $(9x + 8)(3x + 9) = 0$

8) $(4x + 2)(x + 5) = 0$

9) $(x + 2)(2x + 9) = 0$

10) $(12x + 3)(2x + 9) = 0$

11) $2x^2 = 16x$

12) $x^2 - 16 = 0$

13) $2x^2 + 48 = 22x$

14) $-x^2 - 20 = 9x$

15) $x^2 + 8x = 33$

16) $2x^2 + 12x = 80$

17) $x^2 + 14x = -48$

18) $x^2 + 15x = -54$

19) $x^2 + 15x = -36$

20) $x^2 + 2x - 40 = 5x$

21) $x^2 + 16x = -63$

22) $x^2 - 18x = -81$

23) $10x^2 = 7x - 1$

24) $7x^2 - 5x + 8 = 8$

25) $8x^2 + 27 = 33x$

26) $5x^2 - 26x = -24$

27) $3x^2 + 6 = -19x$

28) $x^2 + 22x = -117$

29) $x^2 + 3x - 58 = 30$

30) $5x^2 + 20x - 200 = 25$

31) $3x^2 - 33x + 84 = 0$

32) $6x^2 - 31x + 30 = 15 - 10x^2$

Graphing Quadratic Functions

✎ **Sketch the graph of each function. Identify the vertex and axis of symmetry.**

1) $y = (x + 3)^2 + 5$

2) $y = (x - 3)^2 - 1$

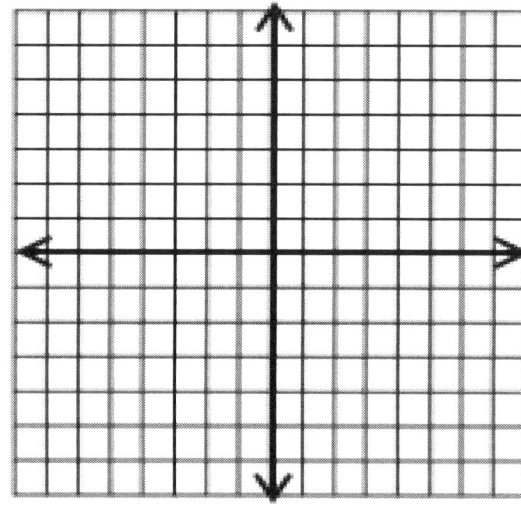
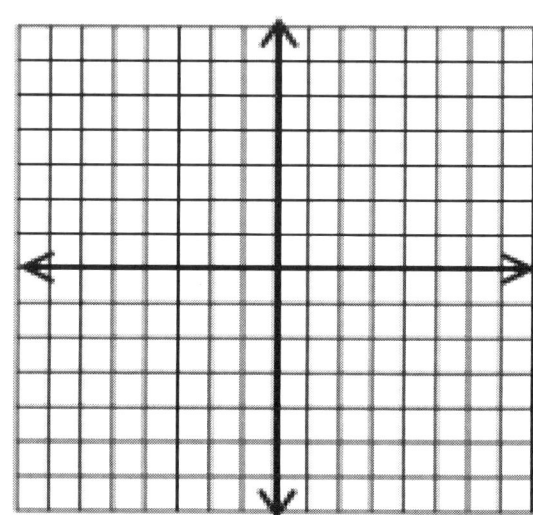

3) $y = 6 - (-x + 2)^2$

4) $y = -3x^2 - 6x + 9$

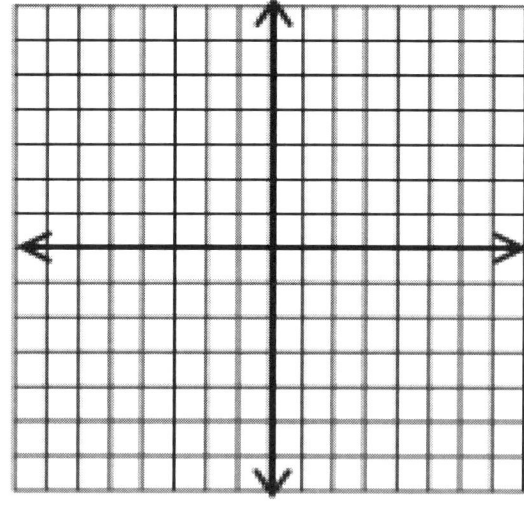
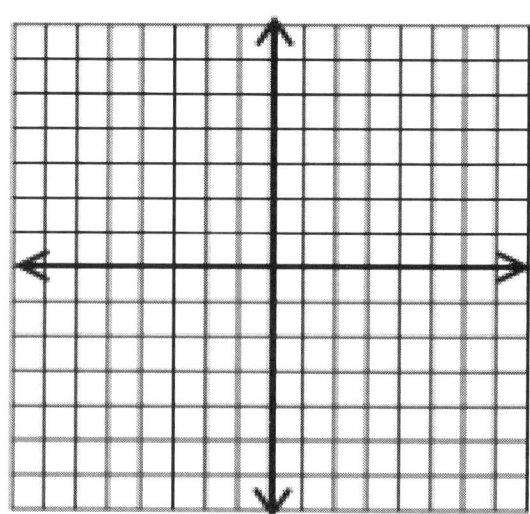

Answers of Worksheets – Chapter 9

Evaluating Function

1) $h(x) = -8x + 9$
2) $k(a) = 5a - 21$
3) $d(t) = 14t$
4) $y(x) = \frac{3}{17}x - \frac{9}{17}$
5) $m(n) = 18n - 94$
6) $c(p) = p^2 - 7p + 15$
7) -25
8) 6.5
9) -1
10) 14
11) -9
12) 38
13) 30
14) -29
15) -3
16) 11.5
17) 2
18) 3
19) 3
20) 12
21) $-\frac{109}{25}$
22) $-\frac{215}{27}$
23) 22
24) -308
25) 5
26) $-\frac{15b+7}{3b}$
27) $8a - 33$
28) $-2x + 7$
29) $2x^2 + 10$
30) $-16x^4 - 8$

Adding and Subtracting Functions

1) 1
2) 2
3) $4t - 6$
4) $-3x^2 - 5x - 7$
5) -26
6) 14
7) 2
8) $a^2 + a + 3$
9) $-8x^2 - 4x - 10$
10) $-z^2 - 4z - 4$
11) $4x^4 - x^2 + 9$
12) $-243x^6 + 12$

Multiplying and Dividing Functions

1) -90
2) -99
3) 180
4) 0
5) $\frac{81}{17}$
6) $\frac{3}{7}$
7) -154
8) -5
9) 0
10) $-\frac{1}{2}$
11) $x^4 - 7x^3 + 12x^2 + 9x - 36$
12) $3x^6 - 17x^4 + 20x^2$

Composition of Functions

1) -5
2) -1
3) 6
4) -2

5) 3

6) -10

7) $\frac{1}{16}$

8) 0

9) 16

10) $\frac{5}{8}$

11) 1

12) 1

13) 5

14) 42

15) -28

16) 2

17) 30

18) $25x - 8$

19) 3

20) -8

21) $3\sqrt{2}$

22) $\sqrt{11}$

23) -5

24) -10

Quadratic Equations

1) $x^2 + 6x - 16$

2) $x^2 + 10x + 9$

3) $x^2 + x - 30$

4) $x^2 + 4x - 21$

5) $x^2 - 17x + 72$

6) $4x^2 - 14x - 8$

7) $3x^2 + 6x - 24$

8) $2x^2 - 11x - 63$

9) $5x^2 - 17x - 12$

10) $12x^2 - 6x - 6$

11) $(x - 7)(x + 3)$

12) $(x + 5)(x + 9)$

13) $(x - 8)(x + 3)$

14) $(x - 1)(x - 6)$

15) $(x + 3)(x + 11)$

16) $(4x + 2)(x + 9)$

17) $(5x - 2)(x + 4)$

18) $(2x - 8)(x + 5)$

19) $(2x + 8)(x + 7)$

20) $4(x - 9)(3x - 10)$

21) $x = -6, x = 9$

22) $x = -1, x = -11$

23) $x = -3$

24) $x = 1, x = -2$

25) $x = 6$

26) $x = -3$

27) $x = -4, x = -\frac{1}{2}$

28) $x = 5, x = -7$

29) $x = 3, x = -9$

30) $x = -1, x = 12$

Solving quadratic equations

1) $\{-5, 2\}$

2) $\{-8, -2\}$

3) $\{9, -5\}$

4) $\{3, 1\}$

5) $\{-9, -4\}$

6) $\{-\frac{5}{2}, -9\}$

7) $\{-\frac{8}{9}, -3\}$

8) $\{-\frac{1}{2}, -5\}$

9) $\{-2, -\frac{9}{2}\}$

10) $\{-\frac{1}{4}, -\frac{9}{2}\}$

11) $\{8, 0\}$

12) $\{4, -4\}$

13) $\{3, 8\}$

14) $\{-5, -4\}$

15) $\{3, -11\}$

16) $\{4, -10\}$

17) $\{-6, -8\}$

18) $\{-6, -9\}$

19) $\{-3, -12\}$

20) $\{8, -5\}$

21) $\{-7, -9\}$

22) $\{9\}$

23) $\{\frac{1}{5}, \frac{1}{2}\}$

24) $\{\frac{5}{7}, 0\}$

25) $\{\frac{9}{8}, 3\}$

26) $\{\frac{6}{5}, 4\}$

27) $\{-\frac{1}{3}, -6\}$

28) $\{-9, -13\}$

29) $\{8, -11\}$

30) $\{5, -9\}$

31) $\{4, 7\}$

32) $\{\frac{15}{16}, 1\}$

Graphing quadratic functions

1) $(-3, 5), x = -3$

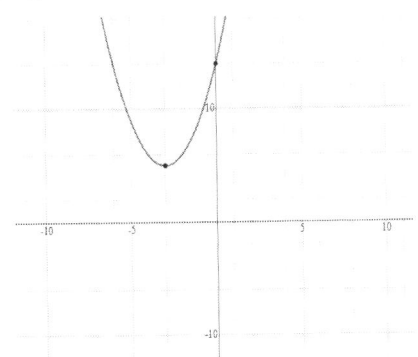

2) $(3, -1), x = 3$

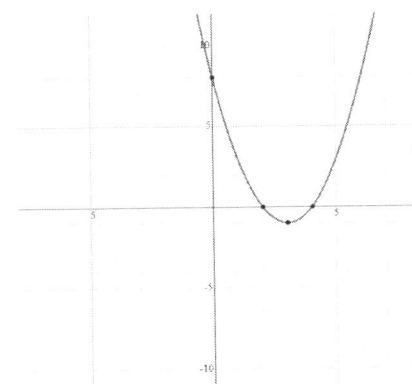

3) $(2, 6), x = 2$

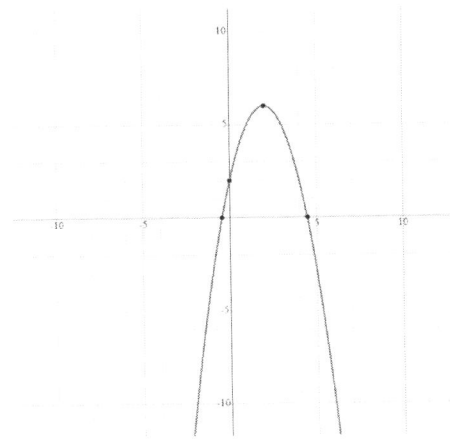

1) $(-1, 12), x = -1$

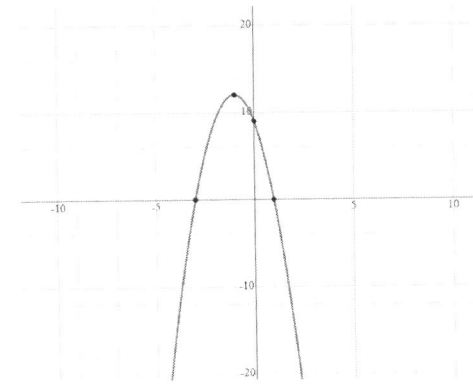

Chapter 10:

Geometry and Solid Figures

Topics that you will practice in this chapter:

- ✓ Angles
- ✓ Pythagorean Relationship
- ✓ Triangles
- ✓ Polygons
- ✓ Trapezoids
- ✓ Circles
- ✓ Cubes
- ✓ Rectangular Prism
- ✓ Cylinder
- ✓ Pyramids and Cone

Mathematics is, as it were, a sensuous logic, and relates to philosophy as do the arts, music, and plastic art to poetry. — *K. Shegel*

Angles

✎ What is the value of x in the following figures?

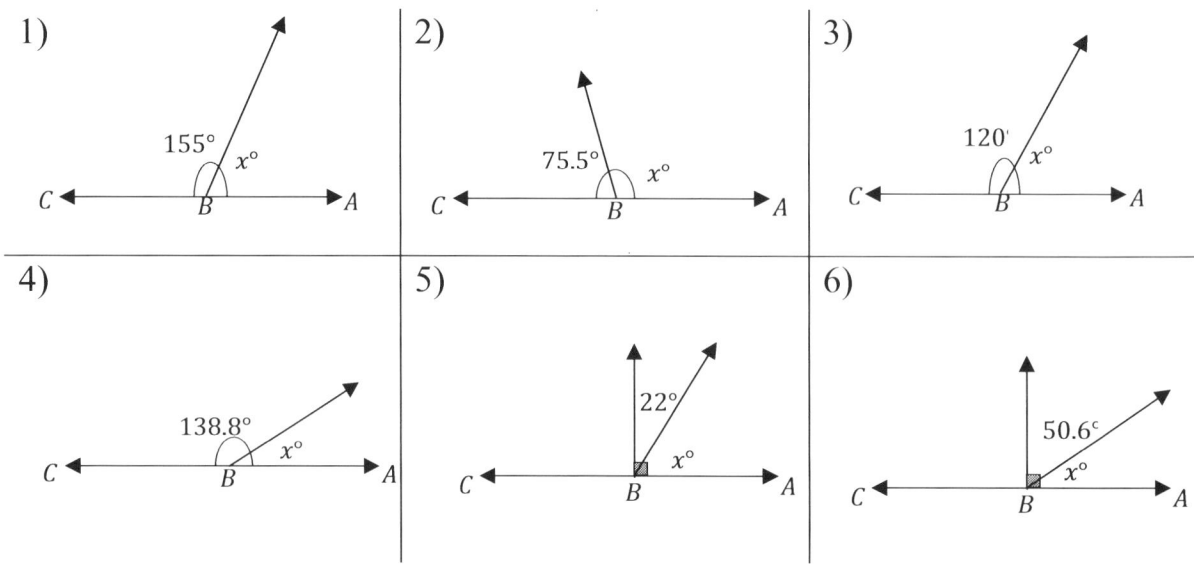

1) $155°$ $x°$ $C \leftarrow B \rightarrow A$

2) $75.5°$ $x°$ $C \leftarrow B \rightarrow A$

3) $120'$ $x°$ $C \leftarrow B \rightarrow A$

4) $138.8°$ $x°$ $C \leftarrow B \rightarrow A$

5) $22°$ $x°$ $C \leftarrow B \rightarrow A$

6) $50.6°$ $x°$ $C \leftarrow B \rightarrow A$

✎ Calculate.

7) Two supplementary angles have equal measures. What is the measure of each angle? _____

8) The measure of an angle is nine seventh the measure of its supplement. What is the measure of the angle? _____

9) Two angles are complementary and the measure of one angle is 24 less than the other. What is the measure of the bigger angle? _____

10) Two angles are complementary. The measure of one angle is one fifth the measure of the other. What is the measure of the smaller angle? _____

11) Two supplementary angles are given. The measure of one angle is 80° less than the measure of the other. What does the bigger angle measure? _____

Pythagorean Relationship

✍ **Do the following lengths form a right triangle?**

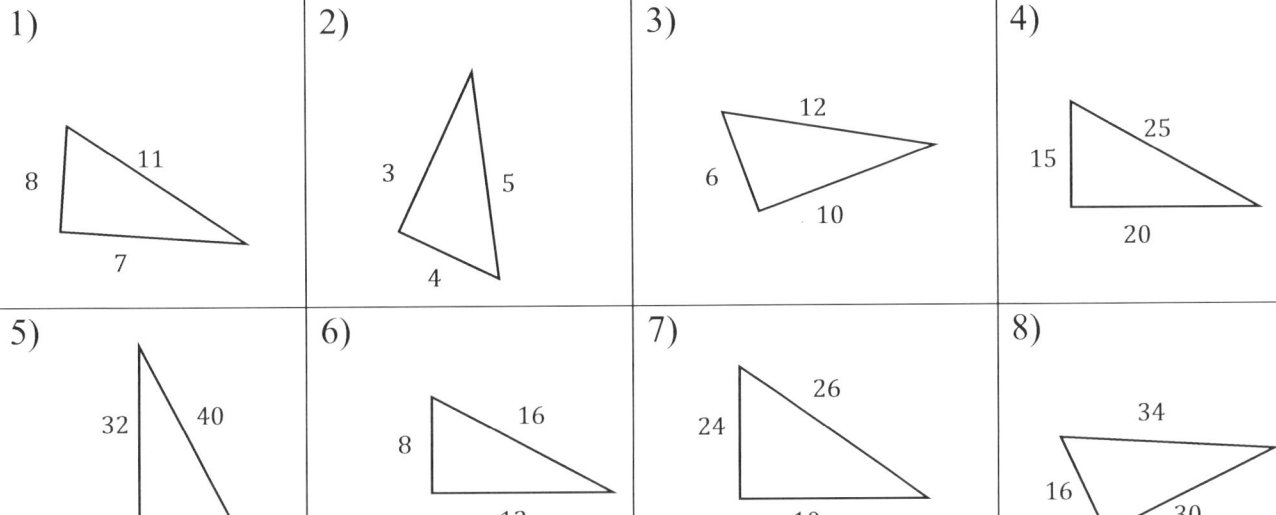

1)

8   11
   7

2)

3   5
   4

3)

   12
6   10

4)

   25
15   20

5)

32   40
   24

6)

8   16
   12

7)

   26
24   10

8)

   34
16   30

✍ **Find the missing side?**

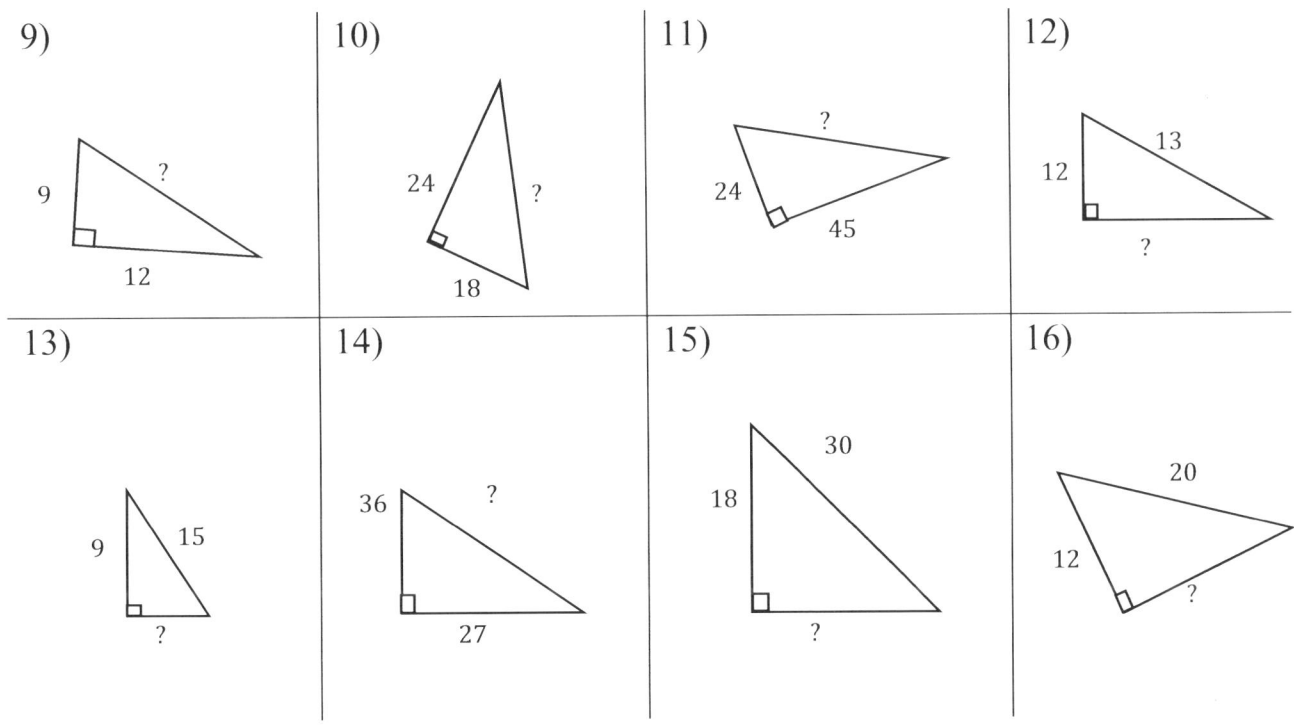

9)

9   ?
   12

10)

24   ?
   18

11)

   ?
24   45

12)

12   13
   ?

13)

9   15
   ?

14)

36   ?
   27

15)

18   30
   ?

16)

   20
12   ?

Triangles

✒ **Find the measure of the unknown angle in each triangle.**

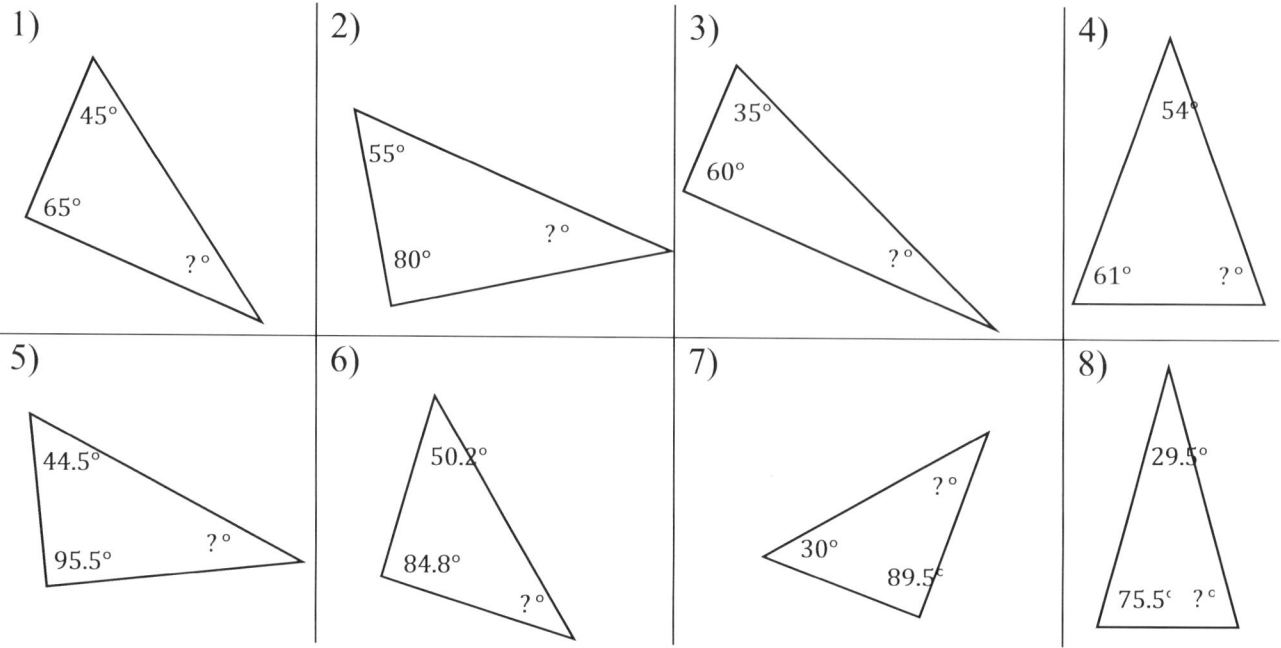

1) 45°, 65°, ?°
2) 55°, 80°, ?°
3) 35°, 60°, ?°
4) 54°, 61°, ?°
5) 44.5°, 95.5°, ?°
6) 50.2°, 84.8°, ?°
7) 30°, 89.5°, ?°
8) 29.5°, 75.5°, ?°

✒ **Find area of each triangle.**

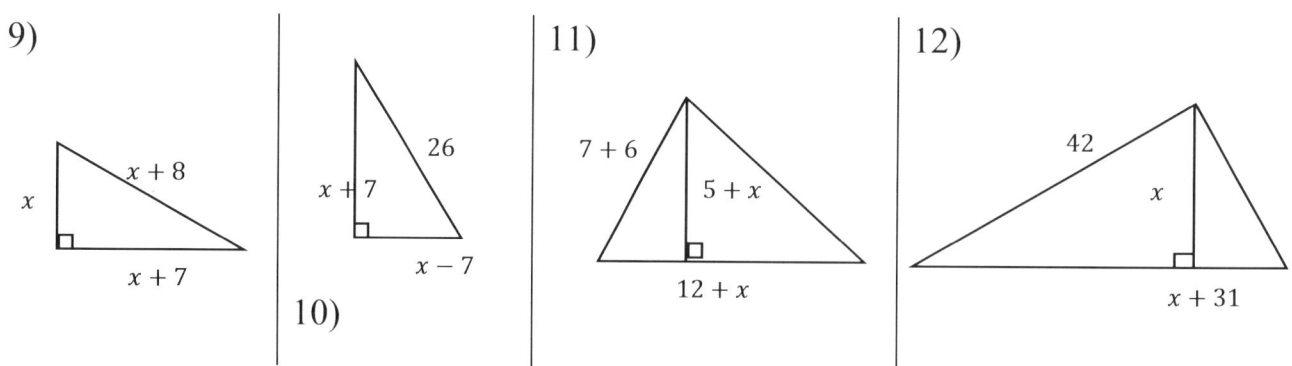

9) x, $x+8$, $x+7$

10) $x+7$, 26, $x-7$

11) $7+6$, $5+x$, $12+x$

12) 42, x, $x+31$

Polygons

✏️ **Find the perimeter of each shape.**

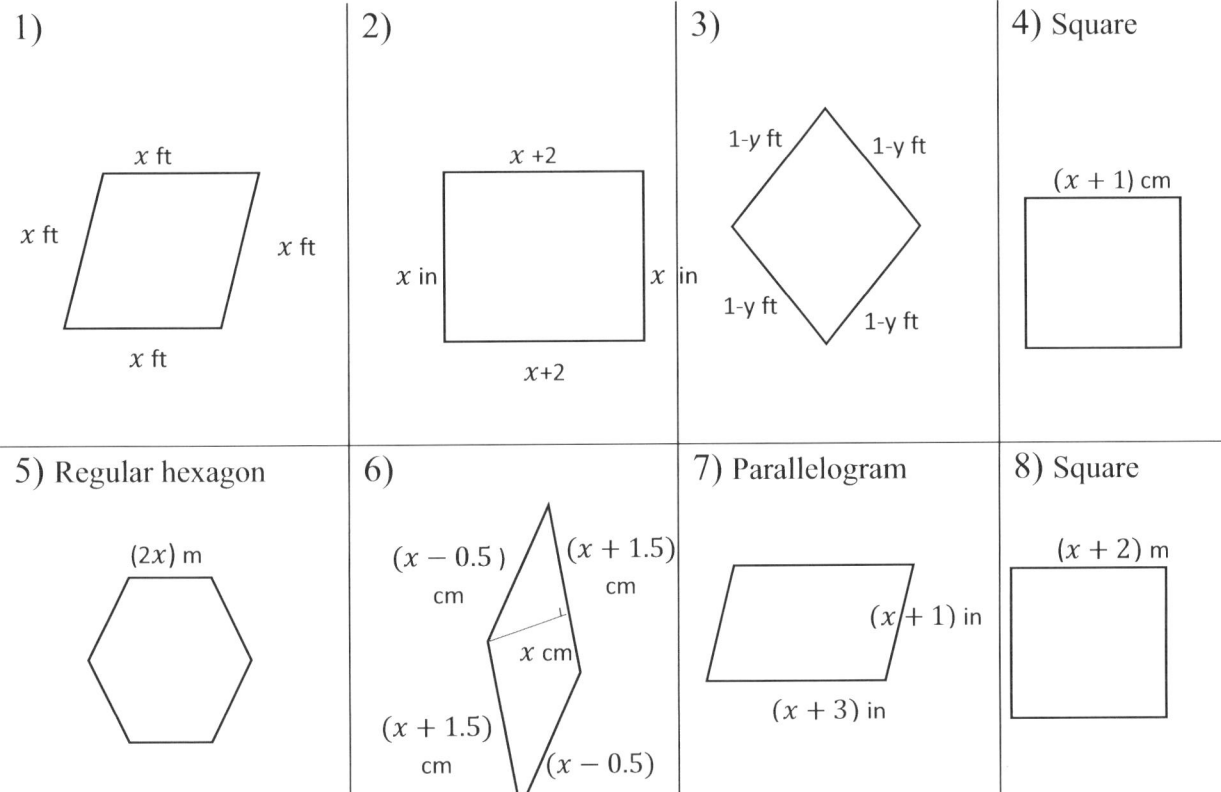

1)

x ft

x ft   x ft

x ft

2)

x +2

x in   x in

x+2

3)

1-y ft   1-y ft

1-y ft   1-y ft

4) Square

(x + 1) cm

5) Regular hexagon

(2x) m

6)

(x − 0.5) cm   (x + 1.5) cm

x cm

(x + 1.5) cm   (x − 0.5) cm

7) Parallelogram

(x + 1) in

(x + 3) in

8) Square

(x + 2) m

✏️ **Find the area of each shape.**

9) Parallelogram

2x m

2x m

2x m

10) Rectangle

10x m²

5x

11) Rectangle

(2+x) km

(2-x) km

12) Square

0.6x m

Trapezoids

✎ **Find the area of each trapezoid.**

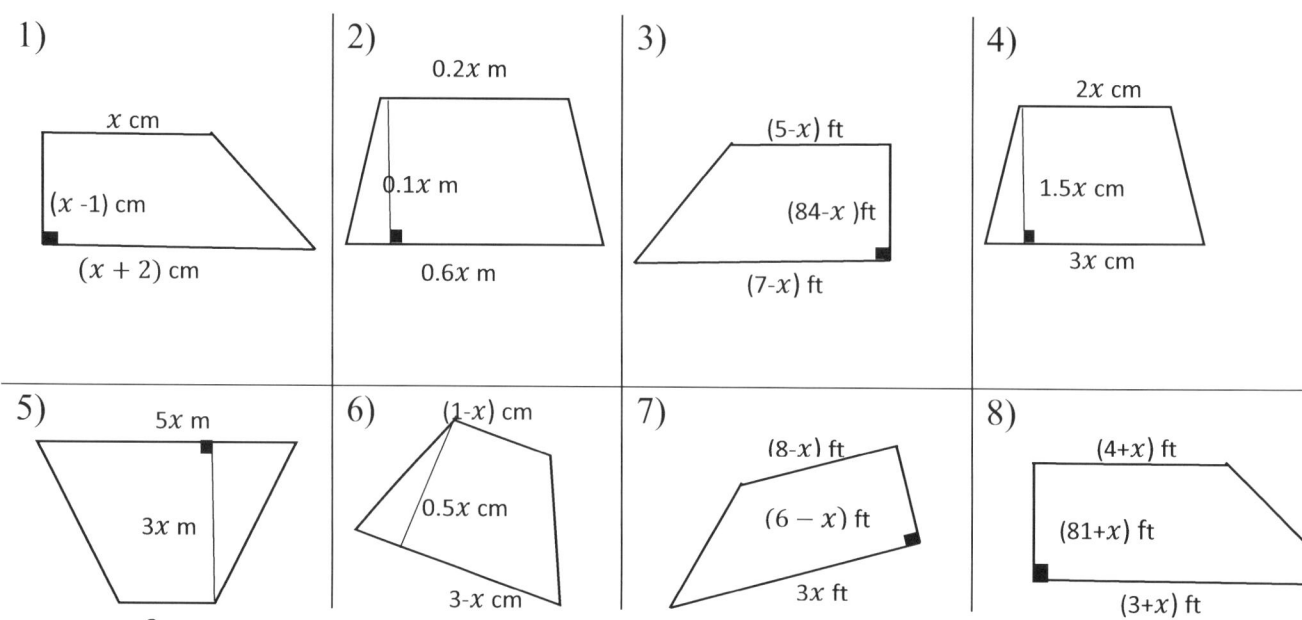

✎ Calculate.

1) A trapezoid has an area of 40 cm² and its height is 8 cm and one base is 6 cm. What is the other base length? _____

2) If a trapezoid has an area of 85 ft² and the lengths of the bases are 9 ft and 8 ft, find the height. _____

3) If a trapezoid has an area of 150 m² and its height is 15 m and one base is 9 m, find the other base length. _____

4) The area of a trapezoid is 196 ft² and its height is 14 ft. If one base of the trapezoid is 12 ft, what is the other base length?

_____

Circles

✎ **Find the area of each circle.** ($\pi = 3.14$)

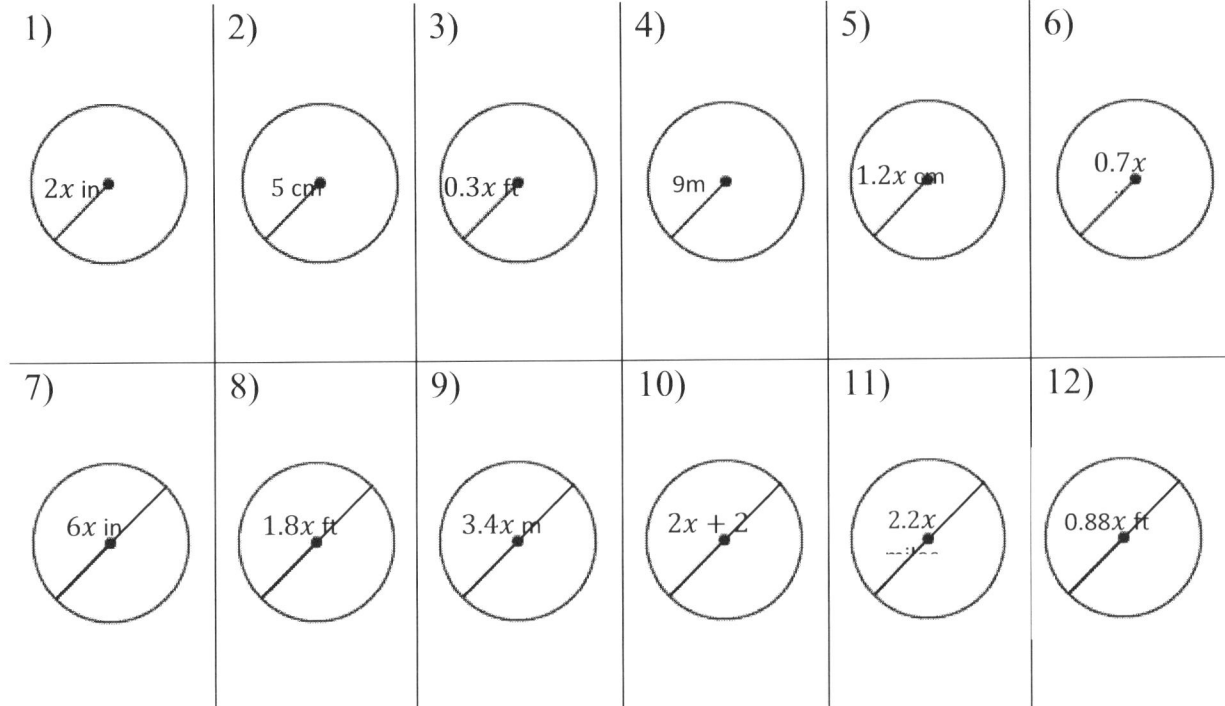

1) $2x$ in

2) 5 cm

3) $0.3x$ ft

4) 9m

5) $1.2x$ cm

6) $0.7x$

7) $6x$ in

8) $1.8x$ ft

9) $3.4x$ m

10) $2x + 2$

11) $2.2x$ miles

12) $0.88x$ ft

✎ **Complete the table below.** ($\pi = 3.14$)

Circle No.	Radius	Diameter	Circumference	Area
1	1.4 inches	2.8 inches	8.792 inches	6.154 square inches
2		4.6 meters		
3				$2.01x^2$ square ft
4			36.42 miles	
5		$6.2x$ kilometers		
6	$5x$ centimeters			
7		$2x$ feet		
8				1.54 square meters
9			$5.7x$ inches	
10	$(1-x)$ feet			

Cubes

✎ **Find the volume of each cube.**

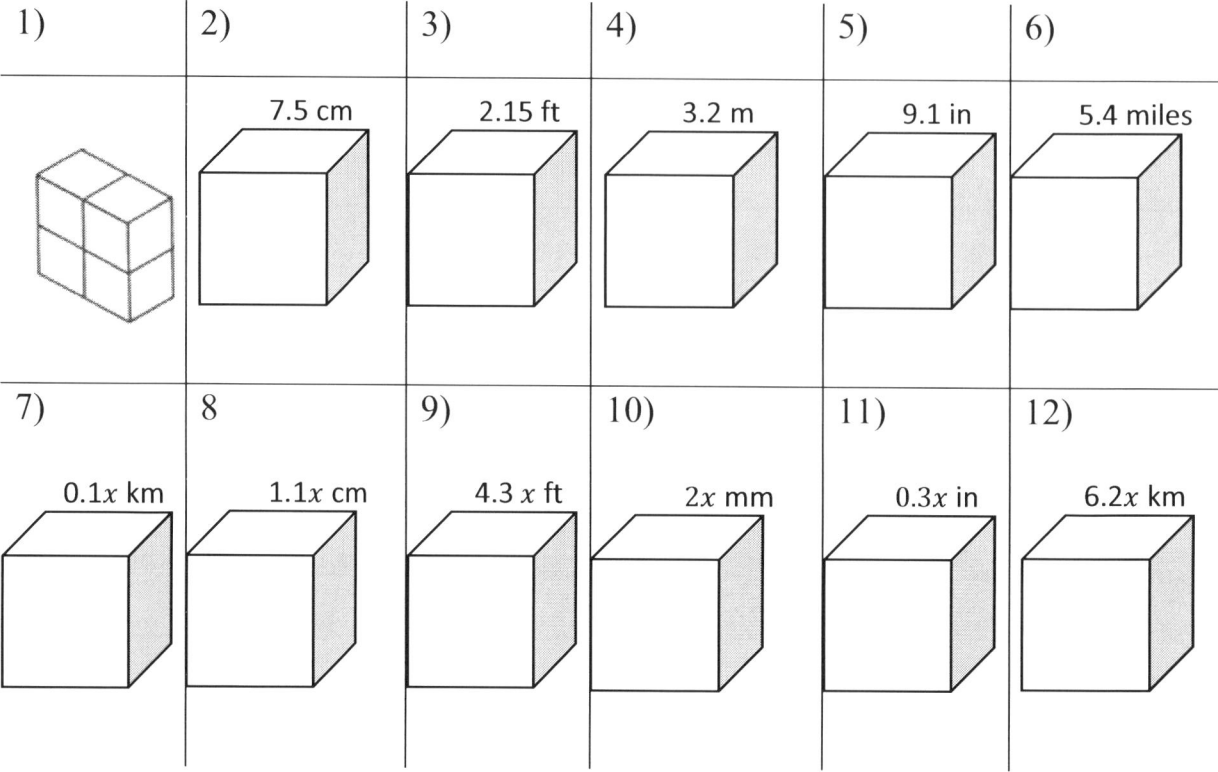

| 1) | 2) 7.5 cm | 3) 2.15 ft | 4) 3.2 m | 5) 9.1 in | 6) 5.4 miles |
| 7) 0.1x km | 8 1.1x cm | 9) 4.3 x ft | 10) 2x mm | 11) 0.3x in | 12) 6.2x km |

✎ **Find the surface area of each cube.**

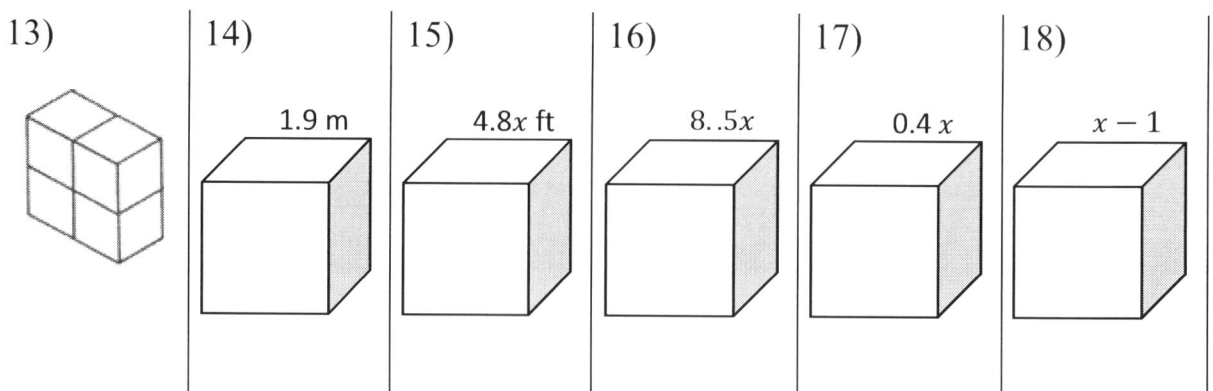

| 13) | 14) 1.9 m | 15) 4.8x ft | 16) 8..5x | 17) 0.4 x | 18) $x-1$ |

Rectangular Prism

✎ **Find the volume of each Rectangular Prism.**

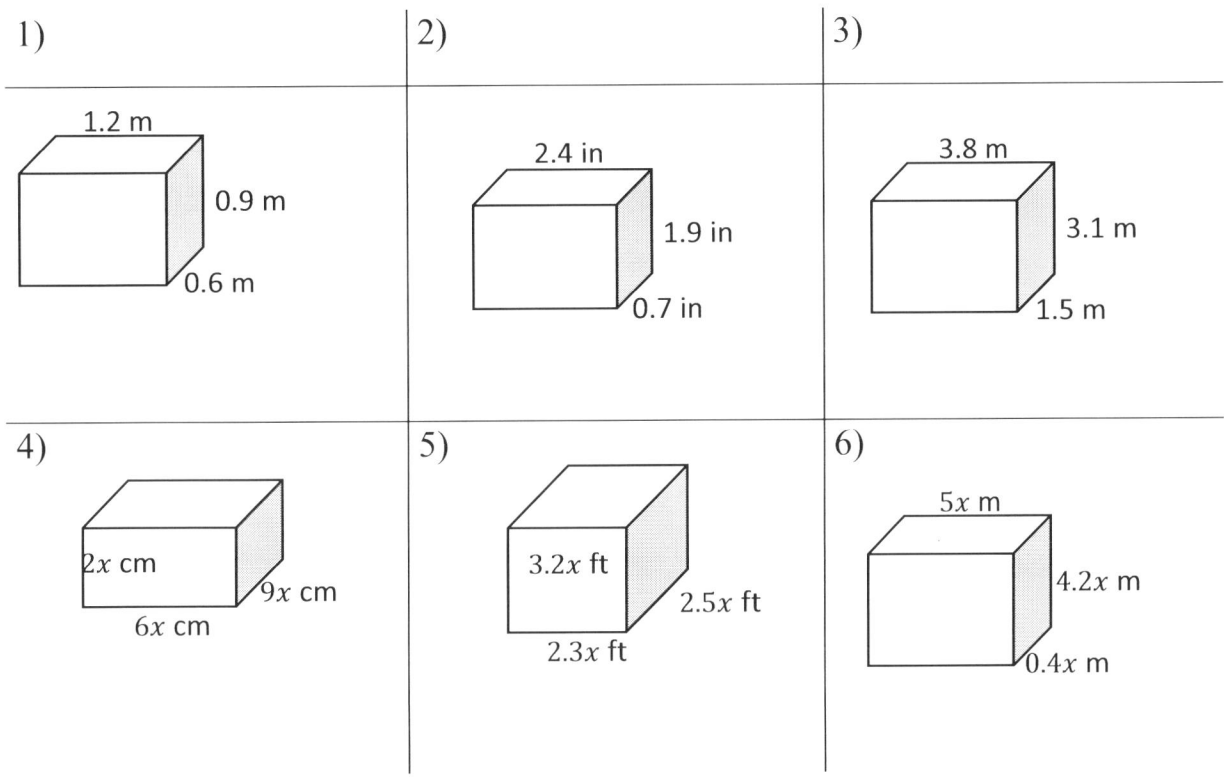

1)

1.2 m
0.9 m
0.6 m

2)

2.4 in
1.9 in
0.7 in

3)

3.8 m
3.1 m
1.5 m

4)

2x cm
9x cm
6x cm

5)

3.2x ft
2.5x ft
2.3x ft

6)

5x m
4.2x m
0.4x m

✎ **Find the surface area of each Rectangular Prism.**

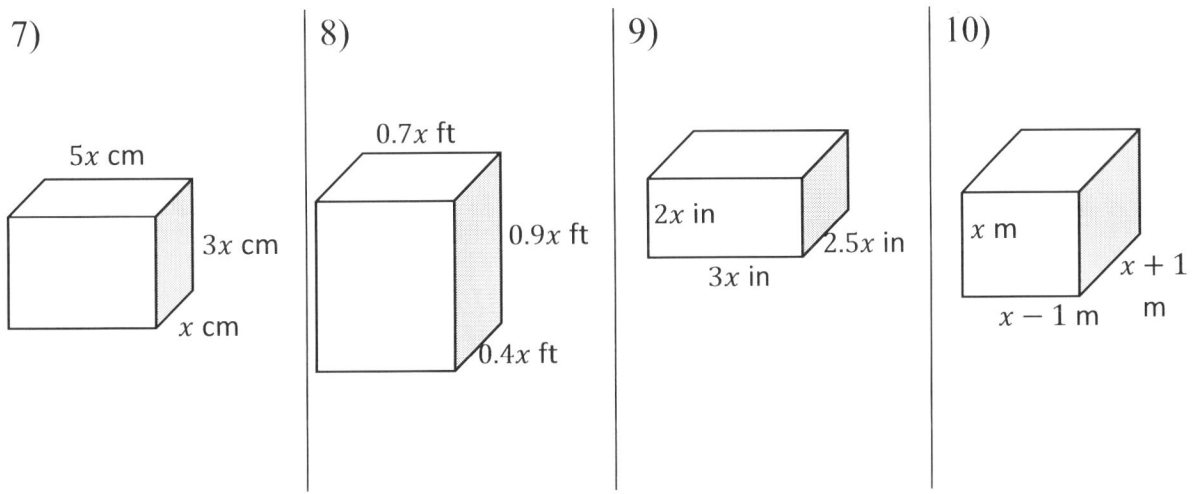

7)

5x cm
3x cm
x cm

8)

0.7x ft
0.9x ft
0.4x ft

9)

2x in
2.5x in
3x in

10)

x m
$x + 1$ m
$x - 1$ m

Cylinder

✍ **Find the volume of each Cylinder. Round your answer to the nearest tenth.** ($\pi = 3.14$)

1)
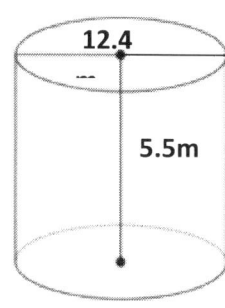
12.4

5.5m

2)
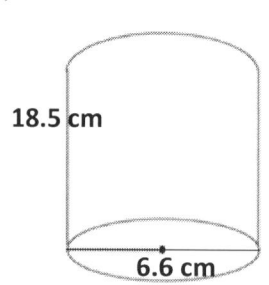
18.5 cm

6.6 cm

3)
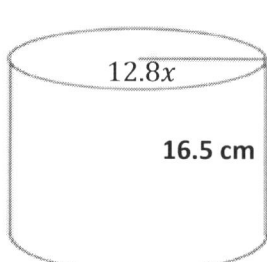
12.8x

16.5 cm

4)
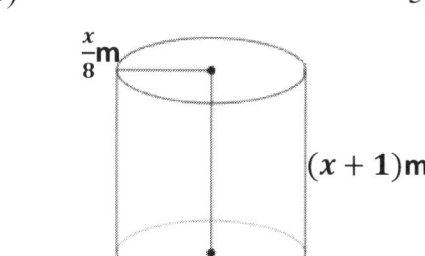
$\frac{x}{8}$m

$(x+1)$m

5)
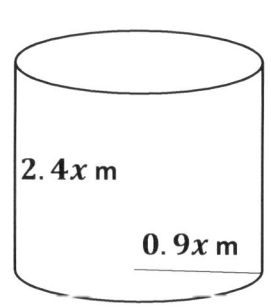
2.4x m

0.9x m

6)
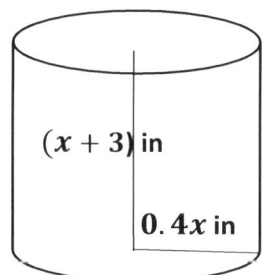
$(x+3)$ in

0.4x in

✍ **Find the surface area of each Cylinder.** ($\pi = 3.14$)

7)
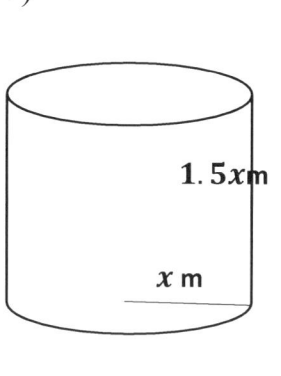
1.5xm

x m

8)
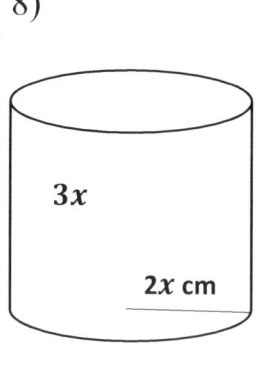
3x

2x cm

9)
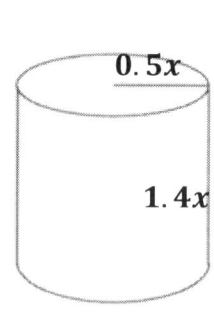
0.5x

1.4x

10)
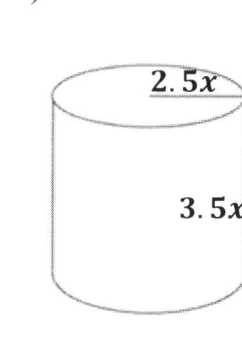
2.5x

3.5x

Pyramids and Cone

✎ **Find the volume of each Pyramid and Cone.** ($\pi = 3.14$)

1)

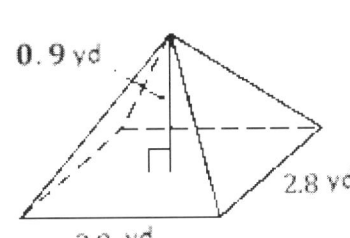

0.9 yd

2.8 yd

2.8 yd

2)

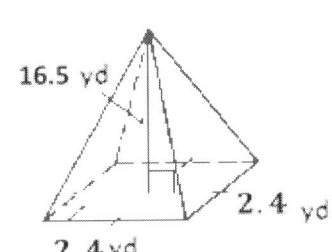

16.5 yd

2.4 yd

2.4 yd

3)

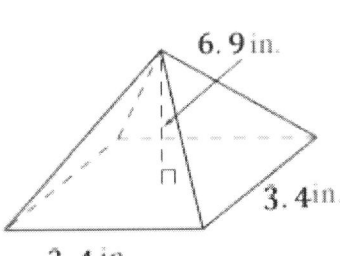

6.9 in.

3.4 in.

3.4 in.

4)

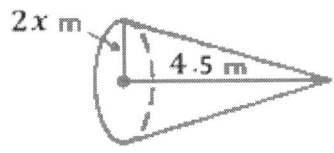

2x m

4.5 m

5)

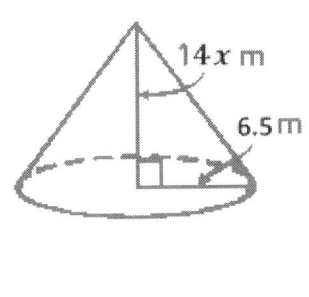

14x m

6.5 m

6)

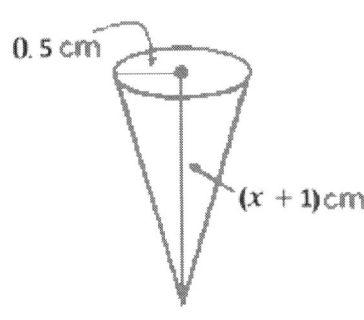

0.5 cm

($x + 1$) cm

✎ **Find the surface area of each Pyramid and Cone.** ($\pi = 3.14$)

7)

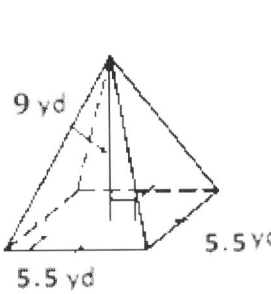

9 yd

5.5 yd

5.5 yd

8)

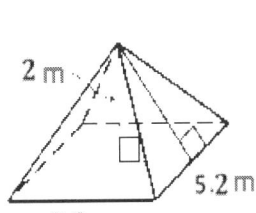

2 m

5.2 m

5.2 m

9)

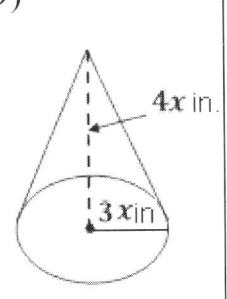

4x in.

3x in

10)

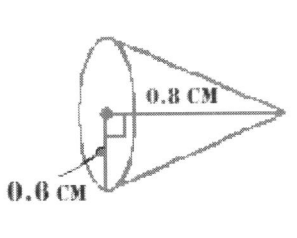

0.8 CM

0.6 CM

Answers of Worksheets – Chapter 10

Angles

1) 25°
2) 104.5°
3) 60°
4) 41.2°
5) 68°
6) 39.4°
7) 90°
8) 101.25°
9) 57°
10) 15°
11) 130°

Pythagorean Relationship

1) *No*
2) *Yes*
3) *No*
4) *Yes*
5) *Yes*
6) *No*
7) *Yes*
8) *Yes*
9) 15
10) 30
11) 51
12) 5
13) 12
14) 45
15) 24
16) 16

Triangles

1) 70°
2) 45°
3) 85°
4) 65°
5) 40°
6) 45°
7) 60.5°
8) 75°
9) $(\frac{x^2+7x}{2})$ square unites
10) $(\frac{x^2-49}{2})$ square unites
11) $(\frac{x^2+17x+60}{2})$ square unites
12) $(\frac{x^2+31x}{2})$ *square unites*

Polygons

1) $(4x)\, ft$
2) $(4x+4)\, in$
3) $(4-4y)\, ft$
4) $(4x+4)\, cm$
5) $(12x)\, m$
6) $(4x+2)\, cm$
7) $(4x+8)\, in$
8) $(4x+8)\, m$
9) $(4x^2) m^2$
10) $(50x^2) m^2$
11) $(4-x^2)\, km^2$
12) $(0.36x^2)\, m^2$

Trapezoids

1) $(x^2-1)\, cm^2$
2) $(0.04x^2)\, m^2$
3) $(x^2-10x+24)\, ft^2$
4) $(3.75x^2)\, cm^2$
5) $(10.5x^2) m^2$
6) $(x-0.5x^2) cm^2$
7) $(-x^2+2x+24)\, ft^2$
8) $(\frac{2x^2+9x+7}{2}) ft^2$

Calculate

1) 4 cm
2) 10 ft
3) 11 m
4) 16 ft

Circles

1) $(12.56x^2)\, in^2$
2) $78.5\, cm^2$
3) $(0.283x^2)\, ft^2$
4) $254.34 m^2$
5) $(4.522x^2) cm^2$
6) $(1.54x^2)\, miles^2$
7) $(28.56x^2)\, in^2$
8) $(2.543x^2) ft^2$
9) $(9.075x^2)\, m^2$

10) $(3.14x^2 + 6.28x + 3.14)\ cm^2$ 11) $(3.8x^2)\ miles^2$ 12) $(0.608x^2)\ ft^2$

Circle No.	Radius	Diameter	Circumference	Area
1	1.4 inches	2.8 inches	8.792 inches	6.154 square inches
2	2.3 meters	4.6 meters	14.44 meters	16.61 meters
3	$0.8x$ square ft	$1.6x$ square ft	$5.024x$ square ft	$2.01x^2$ square ft
4	5.8 miles	11.6 miles	36.42 miles	105.63 miles
5	$3.1x$ kilometers	$6.2x$ kilometers	$19.47x$ kilometers	$30.175x^2$ kilometers
6	$5x$ centimeters	$10x$ centimeters	$31.4x$ centimeters	$78.5x^2$ centimeters
7	x feet	$2x$ feet	$6.28x$ feet	$3.14x^2$ feet
8	0.7 square meters	1.4 square meters	4.396 square meters	1.54 square meters
9	$2.5x$ inches	$5x$ inches	$15.7x$ inches	$19.625x^2$ inches
10	$(1-x)$ feet	$2 - 2x$ feet	$6.28 - 6.28x)$ feet	$3.14x^2 - 6.28x + 3.14)$ feet

Cubes

1) 4
2) $421.88\ cm^3$
3) $9.94\ ft^3$
4) $32.77\ m^3$
5) $753.57\ in^3$

6) $157.46\ miles^3$
7) $(0.001x^3)\ km^3$
8) $(1.33x^3)\ cm^3$
9) $(79.51x^3)\ ft^3$
10) $(8x^3)\ mm^3$

11) $(0.027x^3)\ in^3$
12) $(238.33x^3)\ km^3$
13) 12
14) $21.66\ m^2$
15) $(138.24x^2)\ ft^2$

16) $(433.5x^2)\ mm^2$
17) $(0.96x^2)\ km^2$
18) $6x^2 - 12x + 6\ cm^2$

Rectangular Prism

1) $0.65\ m^3$
2) $3.19\ in^3$
3) $17.67\ m^3$

4) $(108x^3)\ cm^3$
5) $(18.4x^3)\ ft^3$
6) $(8.4x^3)\ m^3$

7) $(46x^2)\ cm^2$
8) $(2.54x^2)\ ft^2$
9) $(37x^2)\ in^2$

10) $(6x^2 - 2)\ m^2$

Cylinder

1) $663.86\ m^3$
2) $632.6\ cm^3$
3) $(8,488.55x^2)\ cm^3$

4) $(0.05x^3 + 0.05x^2)\ m^3$
5) $(6.104x^3)\ m^3$

6) $(0.5\ x^3 + 1.51x^2)in^3$
7) $(15.7x^2)\ m^2$

8) $(62.8x^2)\ cm^2$
9) $(5.97x^2)\ cm^2$
10) $(94.2x^2)\ m^2$

Pyramids and Cone

1) $2.35\ yd^3$
2) $31.68\ yd^3$
3) $26.59\ in^3$

4) $(18.84x^2)\ m^3$
5) $(619.10x)\ m^3$
6) $(0.262x +$

0.262) cm^3
7) $133.77 yd^2$
8) $61.15\ m^2$

9) $(75.36x^2)in^2$
10) $(3.01)cm^2$

Chapter 11:

Statistics and Probability

Topics that you will practice in this chapter:

- ✓ Mean and Median
- ✓ Mode and Range
- ✓ Histograms
- ✓ Stem–and–Leaf Plot
- ✓ Pie Graph
- ✓ Probability Problems
- ✓ Factorials
- ✓ Combinations and Permutation

Mathematics is no more computation than typing is literature.

– John Allen Paulos

Mean and Median

✎ Find Mean and Median of the Given Data.

1) 8, 9, 19, 3, 4

2) 11, 7, 35, 10, 17, 32, 24

3) 38, 9, 15, 17, 13

4) 50, 19, 2, 18, 6, 7

5) 25, 27, 13, 16, 6, 13, 54

6) 24, 364, 42, 57, 6, 68

7) 89, 98, 65, 45, 3, 4, 30, 42

8) 34, 15, 15, 17, 22, 29, 15

9) 2, 5, 10, 45, 8, 13, 35, 6

10) 20, 22, 18, 7, 2, 17, 44, 53

11) 33, 52, 81, 9, 45, 31

12) 19, 74, 51, 8, 12, 15, 9, 14

✎ Calculate.

13) In a javelin throw competition, five athletics score 45, 33, 53, 46 and 19 meters. What are their Mean and Median? _____

14) Eva went to shop and bought 5 apples, 9 peaches, 4 bananas, 7 pineapples and 8 melons. What are the Mean and Median of her purchase? _____

15) Bob has 19 black pen, 15 red pen, 27 green pens, 21 blue pens and one boxes of yellow pens. If the Mean and Median are 19 respectively, what is the number of yellow pens in box? _____

Mode and Range

✎ Find Mode and Rage of the Given Data.

1) 7, 4, 18, 9, 9, 3

   Mode: _____   Range: _____

2) 8, 8, 15, 14, 8, 5, 6, 18

   Mode: _____   Range: _____

3) 4, 4, 4, 15, 19, 24, 31, 5, 4

   Mode: _____   Range: _____

4) 10, 10, 9, 17, 14, 8, 20, 4

   Mode: _____   Range: _____

5) 5, 11, 3, 4, 3, 3

   Mode: _____   Range: _____

6) 13, 7, 7, 7, 7, 4, 12, 25, 8, 3

   Mode: _____   Range: _____

7) 1, 7, 9, 9, 24, 24, 24, 20, 34, 35

   Mode: _____   Range: _____

8) 9, 4, 7, 13, 13, 13, 9, 8, 15

   Mode: _____   Range: _____

9) 8, 8, 8, 5, 8, 7, 17, 16, 3, 9

   Mode: _____   Range: _____

10) 34, 34, 32, 14, 6, 14, 9, 14

   Mode: _____   Range: _____

11) 8, 8, 6, 8, 18, 10, 16, 15

   Mode: _____   Range: _____

12) 12, 12, 7, 11, 14, 12, 33, 5

   Mode: _____   Range: _____

✎ Calculate.

13) A stationery sold 21 pencils, 42 red pens, 25 blue pens, 26 notebooks, 21 erasers, 28 rulers and 27 color pencils. What are the Mode and Range for the stationery sells?

Mode: _____   Range: _____

14) In an English test, eight students score 19, 10, 10, 17, 35, 35, 14 and 10. What are their Mode and Range? _____

15) What is the range of the first 6 odd numbers greater than 8?

_____

Times Series

✎ **Use the following Graph to complete the table.**

Day	Distance (km)
1	
2	

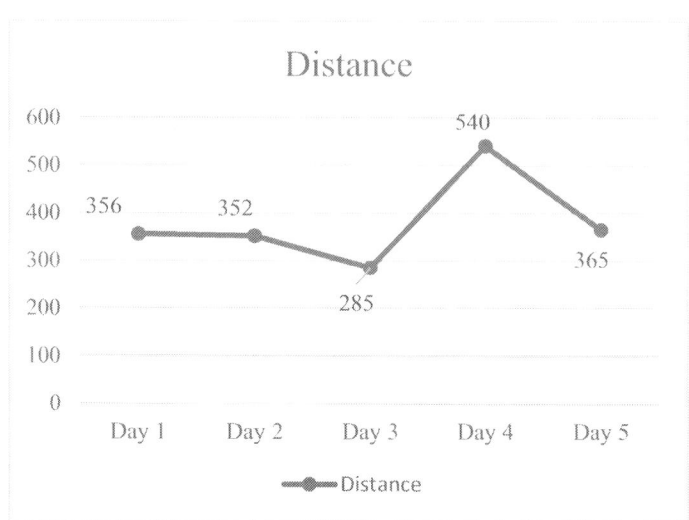

The following table shows the number of births in the US from 2007 to 2012 (in millions).

Year	Number of births (in millions)
2007	4.25
2008	4.19
2009	4.55
2010	3.80
2011	3.25
2012	2.54

Draw a Time Series for the table.

Stem–and–Leaf Plot

✎ **Make stem ad leaf plots for the given data.**

1) 41, 44, 47, 40, 70, 79, 77, 49, 44, 19, 10

Stem	Leaf plot

2) 21, 87, 56, 20, 27, 23, 55, 82, 82, 53, 87, 58

Stem	Leaf plot

3) 111, 47, 66, 44, 94, 117, 62, 114, 48, 112, 68, 99

Stem	Leaf plot

4) 52, 25, 101, 58, 71, 26, 109, 53, 75, 29, 53, 108, 79

Stem	Leaf plot

5) 51, 88, 9, 87, 81, 8, 3, 50, 85, 54, 9, 54, 5

Stem	Leaf plot

6) 40, 93, 20, 25, 48, 92, 95, 52, 21, 44, 97, 29

Stem	Leaf plot

Pie Graph

The circle graph below shows all Robert's expenses for last month. Robert spent $384 on his hobbies last month.

Answer following questions based on the Pie graph.

1) How much was Robert's total expenses last month? _____

2) How much did Robert spend on his car last month? _____

3) How much did Robert spend for shopping last month? _____

4) How much did Robert spend on his rent last month? _____

5) What fraction is Robert's expenses for his car and shopping out of his total expenses last month? _____

Probability Problems

✎ Calculate.

1) A number is chosen at random from 1 to 20. Find the probability of selecting number 8 or smaller numbers. _____

2) Bag A contains 16 red marbles and 6 green marbles. Bag B contains 12 black marbles and 18 orange marbles. What is the probability of selecting a green marble at random from bag A? What is the probability of selecting a black marble at random from Bag B? _____

3) A number is chosen at random from 1 to 25. What is the probability of selecting multiples of 5? _____

4) A card is chosen from a well-shuffled deck of 52 cards. What is the probability that the card will be a queen? _____

5) A number is chosen at random from 1 to 15. What is the probability of selecting a multiple of 4? _____

A spinner, numbered 1–8, is spun once. What is the probability of spinning …?

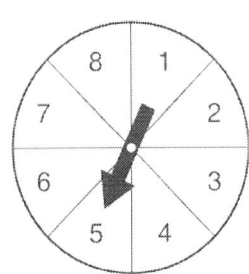

6) an Odd number? _____   7) a multiple of 2? _____

8) a multiple of 5? _____   9) number 10? _____

Factorials

✎ **Determine the value for each expression.**

1) $6! + 1! =$

2) $5! + 2! =$

3) $(4!)^2 =$

4) $6! - 3! =$

5) $8! - 4! + 3 =$

6) $3! \times 4 - 12 =$

7) $(3! + 1!)^2 =$

8) $(5! - 4!)^2 =$

9) $(3! \, 0!)^2 - 2 =$

10) $\dfrac{8!}{6!} =$

11) $\dfrac{3!}{2!} =$

12) $\dfrac{6!}{5!} =$

13) $\dfrac{21!}{19!} =$

14) $\dfrac{(n-1)!}{(n-3)!} =$

15) $\dfrac{(n+2)!}{(n+1)!} =$

16) $\dfrac{(4+2!)^3}{2!} =$

17) $\dfrac{4n!}{2n!} =$

18) $\dfrac{31!}{29!2!} =$

19) $\dfrac{13!}{9!3!} =$

20) $\dfrac{6 \times 280!}{3(4 \times 70)!} =$

21) $\dfrac{30!}{31!2!} =$

22) $\dfrac{7!7!}{8!5!} =$

23) $\dfrac{12!11!}{9!10!} =$

24) $\dfrac{(2 \times 5)!}{1!9!} =$

25) $\dfrac{2!(6n-1)!}{(6n)!} =$

26) $\dfrac{n(4n+4)!}{(4n+5)!} =$

27) $\dfrac{(n+1)!(n)}{(n+2)!} =$

Combinations and Permutations

✎ Calculate the value of each.

1) $6! = $ ____

2) $2! \times 5! = $ ____

3) $4! = $ ____

4) $3! + 5! = $ ____

5) $7! = $ ____

6) $9! = $ ____

7) $3! + 3! = $ ____

8) $5! - 2! = $ ____

✎ Find the answer for each word problems.

9) Susan is baking cookies. She uses sugar, Vanilla and eggs. How many different orders of ingredients can she try? _____

10) Albert is planning for his vacation. He wants to go to museum, watch a movie, go to the beach, play volleyball and play football. How many ways of ordering are there for him? _____

11) How many 6-digit numbers can be named using the digits 1, 6, 8, 9, and 10 without repetition? _____

12) In how many ways can 4 boys be arranged in a straight line? _____

13) In how many ways can 8 athletes be arranged in a straight line? _____

14) A professor is going to arrange her 5 students in a straight line. In how many ways can she do this? _____

15) How many code symbols can be formed with the letters for the word FRIEND? _____

16) In how many ways a team of 7 basketball players can to choose a captain and co-captain? _____

Answers of Worksheets – Chapter 11

Mean and Median

1) Mean: 8.6, Median: 8
2) Mean: 19.43, Median: 17
3) Mean: 18.4, Median: 15
4) Mean: 17, Median: 12.5
5) Mean: 22, Median: 16

6) Mean: 93.5, Median: 49.5
7) Mean: 47, Median: 43.5
8) Mean: 21, Median: 17
9) Mean: 15.5, Median: 9
10) Mean: 22.88, Median: 19

11) Mean: 41.83, Median: 39
12) Mean: 25.25, Median: 14.5
13) Mean: 39.2, Median: 45
14) Mean: 6.6, Median: 7
15) 13

Mode and Range

1) Mode: 9, Range: 15
2) Mode: 8, Range: 13
3) Mode: 4, Range: 27
4) Mode: 10, Range: 16
5) Mode: 3, Range: 8

6) Mode:7, Range: 22
7) Mode:24, Range: 34
8) Mode:13, Range: 11
9) Mode: 8, Range: 14
10) Mode: 14, Range: 28

11) Mode: 8, Range: 12
12) Mode:12, Range: 28
13) Mode: 21, Range: 21
14) Mode: 10, Range: 25
15) 10

Time series

Day	Distance (km)
1	356
2	352
3	285
4	540
5	365

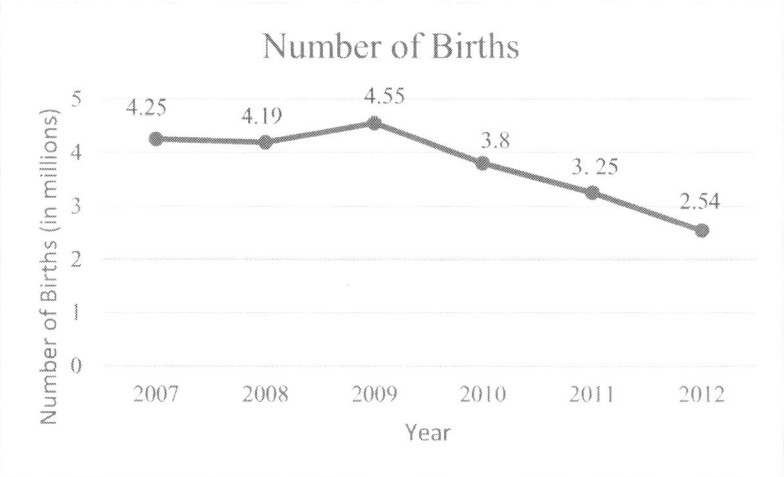

Stem–And–Leaf Plot

1)

Stem	leaf
1	0 9
4	0 1 4 4 5 7 9
7	0 7 9

2)

Stem	leaf
2	0 1 3 7
5	3 5 6 8
8	2 2 7 7

3)

Stem	leaf
4	4 7 8
6	2 6 8
9	4 9
11	1 2 4 7

4)

Stem	leaf
2	5 6 9
5	2 3 3 8
7	1 5 9
10	1 8 9

5)

Stem	leaf
0	3 5 8 9 9
5	0 1 4 4
8	1 5 7 8

6)

Stem	leaf
2	0 1 5 9
4	0 2 4 8
9	2 3 5 7

Pie Graph

1) $1,600

2) $232

3) $312

4) $520

5) $\frac{17}{50}$

Probability Problems

1) $\frac{2}{5}$

2) $\frac{3}{11}, \frac{2}{5}$

3) $\frac{1}{5}$

4) $\frac{1}{13}$

5) $\frac{1}{5}$

6) $\frac{1}{2}$

7) $\frac{1}{2}$

8) $\frac{1}{8}$

9) 0

Factorials

1) 721

2) 122

3) 576

4) 714

5) 40,299

6) 12

7) 49

8) 9,216

9) 34

10) 56

11) 3

12) 6

13) 420

14) $(n-1)(n-2)$

15) $n+2$

16) 108

17) 2

18) 465

19) 2,860

20) 2

21) $\frac{1}{62}$

22) 5.25

23) 14,520

24) 10

25) $\frac{1}{3n}$

26) $\frac{n}{4n+5}$

27) $\frac{n}{n+2}$

Combinations and Permutations

1) 720

2) 240

3) 24

4) 126

5) 5,040

6) 362,880

7) 12

8) 118

9) 6

10) 120

11) 720

12) 24

13) 40,320

14) 120

15) 720

16) 42

ISEE Upper Level Practice Tests

The Independent School Entrance Exam (ISEE) is a standardized test developed by the Educational Records Bureau for its member schools as part of their admission process. There are currently four Levels of the ISEE:

- ✓ Primary Level (entering Grades 2 - 4)
- ✓ Lower Level (entering Grades 5 and 6)
- ✓ Middle Level (entering Grades 7 and 8)
- ✓ Upper Level (entering Grades 9 - 12)

There are five sections on the ISEE Upper Level Test:

- o Verbal Reasoning
- o Quantitative Reasoning
- o Reading Comprehension
- o Mathematics Achievement
- o and a 30-minute essay

ISEE Upper Level tests use a multiple-choice format and contain two Mathematics sections:

Quantitative Reasoning

There are 37 questions in the Quantitative Reasoning section and students have 35 minutes to answer the questions. This section contains word problems and quantitative comparisons. The word problems require either no calculation or simple calculation. The quantitative comparison items present two quantities, (A) and (B), and the student needs to select one of the following four answer choices:

(A) The quantity in Column A is greater.

(B) The quantity in Column B is greater.

(C) The two quantities are equal.

(D) The relationship cannot be determined from the information given.

Mathematics Achievement

There are 47 questions in the Mathematics Achievement section and students have 40 minutes to answer the questions. Mathematics Achievement measures students' knowledge of Mathematics requiring one or more steps in calculating the answer.

In this book, we have reviewed Quantitative Reasoning and Mathematic Achievement topics being tested on the ISEE Upper Level. In this section, there are two complete ISEE Upper Level Quantitative Reasoning and Mathematics Achievement Tests. Let your student take these tests to see what score they will be able to receive on a real ISEE Upper Level test.

Good Luck!

Time to Test

Time to refine your skill with a practice examination

Take a practice ISEE Upper Level Math Test to simulate the test day experience. After you've finished, score your test using the answer key.

Before You Start

- You'll need a pencil and scratch papers to take the test.
- For each question, there are four possible answers. Choose which one is best.
- It's okay to guess. You won't lose any points if you're wrong.
- Use the answer sheet provided to record your answers.
- After you've finished the test, review the answer key to see where you went wrong.
- **Calculators are NOT allowed for the ISEE Upper Level Test.**

Good Luck!

ISEE Upper Level Practice Test Answer Sheets

Remove (or photocopy) these answer sheets and use them to complete the practice tests.

ISEE Upper Level Practice Test

Quantitative Reasoning

1	Ⓐ Ⓑ Ⓒ Ⓓ	21	Ⓐ Ⓑ Ⓒ Ⓓ
2	Ⓐ Ⓑ Ⓒ Ⓓ	22	Ⓐ Ⓑ Ⓒ Ⓓ
3	Ⓐ Ⓑ Ⓒ Ⓓ	23	Ⓐ Ⓑ Ⓒ Ⓓ
4	Ⓐ Ⓑ Ⓒ Ⓓ	24	Ⓐ Ⓑ Ⓒ Ⓓ
5	Ⓐ Ⓑ Ⓒ Ⓓ	25	Ⓐ Ⓑ Ⓒ Ⓓ
6	Ⓐ Ⓑ Ⓒ Ⓓ	26	Ⓐ Ⓑ Ⓒ Ⓓ
7	Ⓐ Ⓑ Ⓒ Ⓓ	27	Ⓐ Ⓑ Ⓒ Ⓓ
8	Ⓐ Ⓑ Ⓒ Ⓓ	28	Ⓐ Ⓑ Ⓒ Ⓓ
9	Ⓐ Ⓑ Ⓒ Ⓓ	29	Ⓐ Ⓑ Ⓒ Ⓓ
10	Ⓐ Ⓑ Ⓒ Ⓓ	30	Ⓐ Ⓑ Ⓒ Ⓓ
11	Ⓐ Ⓑ Ⓒ Ⓓ	31	Ⓐ Ⓑ Ⓒ Ⓓ
12	Ⓐ Ⓑ Ⓒ Ⓓ	32	Ⓐ Ⓑ Ⓒ Ⓓ
13	Ⓐ Ⓑ Ⓒ Ⓓ	33	Ⓐ Ⓑ Ⓒ Ⓓ
14	Ⓐ Ⓑ Ⓒ Ⓓ	34	Ⓐ Ⓑ Ⓒ Ⓓ
15	Ⓐ Ⓑ Ⓒ Ⓓ	35	Ⓐ Ⓑ Ⓒ Ⓓ
16	Ⓐ Ⓑ Ⓒ Ⓓ	36	Ⓐ Ⓑ Ⓒ Ⓓ
17	Ⓐ Ⓑ Ⓒ Ⓓ	37	Ⓐ Ⓑ Ⓒ Ⓓ
18	Ⓐ Ⓑ Ⓒ Ⓓ	38	Ⓐ Ⓑ Ⓒ Ⓓ
19	Ⓐ Ⓑ Ⓒ Ⓓ	39	Ⓐ Ⓑ Ⓒ Ⓓ
20	Ⓐ Ⓑ Ⓒ Ⓓ	40	Ⓐ Ⓑ Ⓒ Ⓓ

Mathematics Achievement

1	Ⓐ Ⓑ Ⓒ Ⓓ	21	Ⓐ Ⓑ Ⓒ Ⓓ	41	Ⓐ Ⓑ Ⓒ Ⓓ
2	Ⓐ Ⓑ Ⓒ Ⓓ	22	Ⓐ Ⓑ Ⓒ Ⓓ	42	Ⓐ Ⓑ Ⓒ Ⓓ
3	Ⓐ Ⓑ Ⓒ Ⓓ	23	Ⓐ Ⓑ Ⓒ Ⓓ	43	Ⓐ Ⓑ Ⓒ Ⓓ
4	Ⓐ Ⓑ Ⓒ Ⓓ	24	Ⓐ Ⓑ Ⓒ Ⓓ	44	Ⓐ Ⓑ Ⓒ Ⓓ
5	Ⓐ Ⓑ Ⓒ Ⓓ	25	Ⓐ Ⓑ Ⓒ Ⓓ	45	Ⓐ Ⓑ Ⓒ Ⓓ
6	Ⓐ Ⓑ Ⓒ Ⓓ	26	Ⓐ Ⓑ Ⓒ Ⓓ	46	Ⓐ Ⓑ Ⓒ Ⓓ
7	Ⓐ Ⓑ Ⓒ Ⓓ	27	Ⓐ Ⓑ Ⓒ Ⓓ	47	Ⓐ Ⓑ Ⓒ Ⓓ
8	Ⓐ Ⓑ Ⓒ Ⓓ	28	Ⓐ Ⓑ Ⓒ Ⓓ	48	Ⓐ Ⓑ Ⓒ Ⓓ
9	Ⓐ Ⓑ Ⓒ Ⓓ	29	Ⓐ Ⓑ Ⓒ Ⓓ	49	Ⓐ Ⓑ Ⓒ Ⓓ
10	Ⓐ Ⓑ Ⓒ Ⓓ	30	Ⓐ Ⓑ Ⓒ Ⓓ	50	Ⓐ Ⓑ Ⓒ Ⓓ
11	Ⓐ Ⓑ Ⓒ Ⓓ	31	Ⓐ Ⓑ Ⓒ Ⓓ		
12	Ⓐ Ⓑ Ⓒ Ⓓ	32	Ⓐ Ⓑ Ⓒ Ⓓ		
13	Ⓐ Ⓑ Ⓒ Ⓓ	33	Ⓐ Ⓑ Ⓒ Ⓓ		
14	Ⓐ Ⓑ Ⓒ Ⓓ	34	Ⓐ Ⓑ Ⓒ Ⓓ		
15	Ⓐ Ⓑ Ⓒ Ⓓ	35	Ⓐ Ⓑ Ⓒ Ⓓ		
16	Ⓐ Ⓑ Ⓒ Ⓓ	36	Ⓐ Ⓑ Ⓒ Ⓓ		
17	Ⓐ Ⓑ Ⓒ Ⓓ	37	Ⓐ Ⓑ Ⓒ Ⓓ		
18	Ⓐ Ⓑ Ⓒ Ⓓ	38	Ⓐ Ⓑ Ⓒ Ⓓ		
19	Ⓐ Ⓑ Ⓒ Ⓓ	39	Ⓐ Ⓑ Ⓒ Ⓓ		
20	Ⓐ Ⓑ Ⓒ Ⓓ	40	Ⓐ Ⓑ Ⓒ Ⓓ		

ISEE Upper Level Practice Test 1

Mathematics

Quantitative Reasoning

❖ **37 Questions.**

❖ **Total time for this test: 35 Minutes.**

❖ **Calculators are not allowed at the test.**

Administered *Month Year*

1) Which of the following is NOT a factor of 80?

    A. 8   C. 16

    B. 20   D. 14

2) There are 9 blue marbles, 10 red marbles, and 6 yellow marbles in a box. If Mia randomly selects a marble from the box, what is the probability of selecting a blue or yellow marble?

    A. $\frac{4}{25}$ C. $\frac{3}{5}$

    B. $\frac{2}{15}$ D. $\frac{3}{10}$

3) On a map, the length of the road from City A to City B is measured to be 18 inches. On this map, $\frac{1}{4}$ inch represents an actual distance of 6 miles. What is the actual distance, in miles, from City A to City B along this road?

    A. 42   C. 32

    B. 432   D. 324

4) What is the area of a square whose diagonal is 12?

    A. 12   C. 72

    B. 24   D. 56

5) If Emily left a $23.75 tip on a breakfast that cost $58.25, approximately what percentage was the tip?

    A. 14%   C. 41%

    B. 28%   D. 52%

6) A phone company charges $6 for the first three minutes of a phone call and 30 cents per minute thereafter. If Sofia makes a phone call that lasts 20 minutes, what will be the total cost of the phone call?

A. 10.10

C. 11.10

B. 9.90

D. 12

7) Michelle and Alec can finish a job together in 90 minutes. If Michelle can do the job by herself in 3 hours, how many minutes does it take Alec to finish the job?

A. 160

C. 720

B. 180

D. 280

8) Elise earns $6.50 per hour and worked 24 hours. Bob earns $6.00 per hour. How many hours would Bob need to work to equal James's earnings over 24 hours?

A. 36

C. 16

B. 60

D. 26

9) In the figure, MN is 48 cm. How long is ON?

A. 12 cm

B. 32 cm

C. 16 cm

D. 42 cm

Use following graph to answer questions 10 and 11.

A library has 400 books that include Mathematics, Physics, Chemistry, English and History.

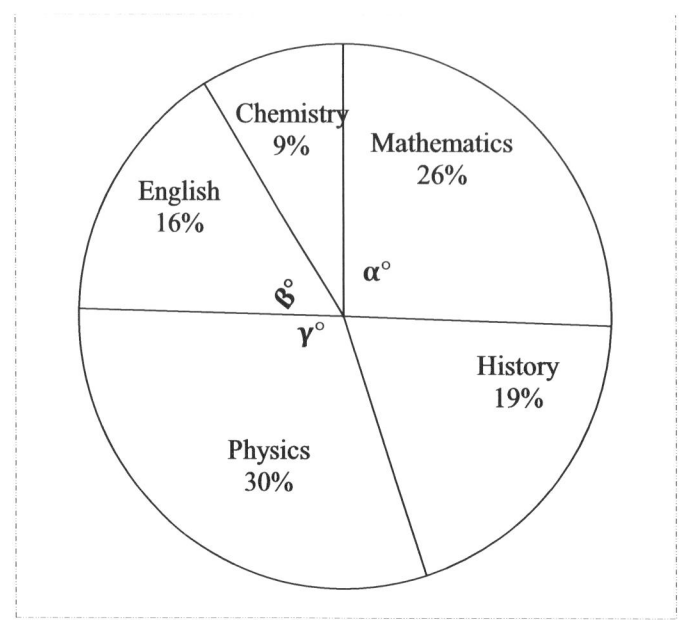

10) What is the product of the number of Physics and number of Chemistry books?

A. 4,120   C. 8,750

B. 4,320   D. 8,640

11) If 160% of a number is 48, then what is the 40% of that number?

A. 12   C. 25

B. 8   D. 40

12) If $5y + 6 < 41$, then y could be equal to?

A. 12   C. 10.5

B. 16   D. 6.5

13) The first five terms in a sequence are shown below. What is the seventh term in the sequence?

$$\{3, 6, 11, 18, 27, \ldots\}$$

A.52
C.46

B.39
D.37

14) Emily and Daniel have taken the same number of photos on their school trip. Emily has taken 8 times as many as photos as Claire and Daniel has taken 21 more photos than Claire. How many photos has Claire taken?

A. 3
C. 8

B. 4
D. 7

15) What is the equation of the line that passes through $(5, -2)$ and has a slope of 2?

A.$y = 2x - 12$
C.$y = -2x + 10$

B.$y = -2x - 12$
D.$y = 2x + 10$

16) A supermarket's sales increased by 30 percent in the month of April and decreased by 30 percent in the month of May. What is the percent change in the sales of the supermarket over the two-month period?

A.3% increase
C.9% decrease

B.No change
D.6% decrease

17) How many $\frac{1}{8}$ pound paperback books together weigh 12 pounds?

A.9.6
C.96

B.19
D.106

18) Find the solution (x, y) to the following system of equations?

$$-3x + y = 5$$

$$-6x + 4y = 14$$

A. $(2, 6)$ C. $(2, -1)$

B. $(-6, 2)$ D. $(-1, 2)$

19) The distance between cities A and B is approximately 4,800 miles. If Alice drives an average of 76 miles per hour, how many hours will it take Alice to drive from city A to city B?

A. Approximately 63 hours   C. Approximately 78 hours

B. Approximately 68 hours   D. Approximately 81 hours

20) Three seventh of 28 is equal to $\frac{4}{9}$ of what number?

A. 32   C. 54

B. 16   D. 27

21) Sophia purchased a sofa for $496.40. The sofa is regularly priced at $584. What was the percent discount Sophia received on the sofa?

A. 15%   C. 25%

B. 5%   D. 55%

Quantitative Comparisons

Direction: Questions 22 to 37 are Quantitative Comparisons Questions. Using the information provided in each question, compare the quantity in column A to the quantity in Column B. Choose on your answer sheet grid

A   if the quantity in Column A is greater

B   if the quantity in Column B is greater

C   if the two quantities are equal

D   if the relationship cannot be determined from the information given

22) Set A includes odd primes numbers less than 10.

Column A	**Column B**
The sum of all members in Set A	18

23)

Column A	**Column B**
1.8%	$\dfrac{1}{6}$

24)

Column A	**Column B**
The average of 11, 14, 19, 21, 34	The average of 22, 24, 34, 28

25)

Column A	**Column B**
$8 + 12(9 - 8)$	$12 + 8(9 - 8)$

26)

Column A	Column B
The number of posts needed for a fence 138 feet long if the posts are placed 11.5 feet apart	15 posts

27) $\frac{x}{56} = \frac{3}{14}$

Column A	Column B
$\frac{6}{x}$	$\frac{1}{2}$

28) Working at constant rates, machine D makes b rolls of steel in 15 minutes and machine E makes b rolls of steel in 12 minutes $(b > 0)$

Column A	Column B
The number of rolls of steel made by machine D in 1 hours and 45 minutes.	The number of rolls of steel made by machine E in 1 hours.

29) The ratio of boys to girls in a class is 10 to 12.

Column A	Column B
Ratio of girls to the entire class	$\frac{1}{5}$

30) There are 8 blue marbles and 6 green marbles in a jar. Two marbles are pulled out in succession without replacing them in the jar.

Column A	**Column B**
The probability that both marbles are blue.	The probability that the first marbles is green, but the second is blue.

31) $\frac{x}{9} = y^2$

Column A	**Column B**
x	y

32) $x = 1$

Column A	**Column B**
$5x^2 - 4x + 14$	$3x^3 + 2x^2 + 8$

33) $4 > y > -2$

Column A	**Column B**
$\dfrac{y}{5}$	$\dfrac{5}{y}$

34) $\frac{a}{c} = \frac{d}{b}$

Column A	**Column B**
$c - a$	$b - d$

35) A magazine printer consecutively numbered the pages of a magazine, starting with 1 on the first page, 12 on the twelfth page, etc. In numbering the pages, the printer printed a total of 213 digits.

Column A	Column B
The number of pages in the magazine	106

36) A computer priced $126 includes 5% profit

Column A	Column B
$115	The original cost of the computer

37)

Column A	Column B
The largest number that can be written by rearranging the digits in 295	The largest number that can be written by rearranging the digits in 674

STOP

IF YOU FINISH BEFORE TIME IS CALLED, YOU MAY CHECK YOUR WORK ON THIS SECTION ONLY. DO NOT TURN TO ANY OTHER SECTION IN THE TEST.

ISEE Upper Level Practice Test 1

Mathematics

Mathematics Achievement

❖ **47 Questions.**

❖ **Total time for this test: 40 Minutes**.

❖ **Calculators are not allowed at the test.**

Administered *Month Year*

1) $8 - 14 \div (7^2 \div 7) =$ ___

    A. 6   C. −2

    B. −6   D. 2

2) How is this number written in scientific notation?

$$0.0002796$$

    A. 2.796×10^4 C. $2,796 \times 10^6$

    B. 2.796×10^{-4} D. $2,7963 \times 10^{-6}$

3) A girl 120 cm tall, stands 160 cm from a lamp post at night. Her shadow from the light is 40 cm long. How high is the lamp post?

    A. 200

    B. 1,000

    C. 600

    D. 400

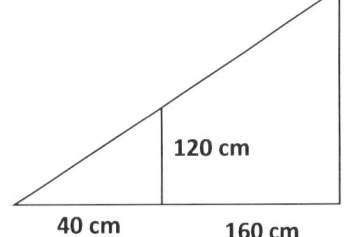

4) Which value of x makes the following inequality true?

$$\frac{7}{31} \le x < 25\%$$

    A. 0.27   C. $\sqrt{0.0169}$

    B. $\frac{15}{65}$ D. $(0.26)^2$

5) $(x + 7)(x + 9) =$

    A. $x^2 + 63x + 16$ C. $x^2 + 16x + 63$

    B. $2x - 16x - 63$ D. $x^2 - 16x + 63$

6) Which of the following graphs represents the compound inequality $-3 \leq 2x - 5 < 7$?

A.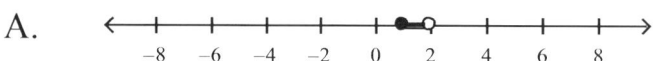

B.

C.

D.

7) Find all values of x for which $6x^2 + 21x + 9 = 0$

A. $-\frac{1}{3}, -2$

B. $-\frac{1}{2}, -3$

C. $2, \frac{1}{3}$

D. $-4, \frac{1}{3}$

8) Emily lives $8\frac{1}{8}$ miles from where she works. When traveling to work, she walks to a bus stop $\frac{1}{5}$ of the way to catch a bus. How many miles away from her house is the bus stop?

A. $2\frac{1}{8}$ Miles

B. $1\frac{5}{16}$ Miles

C. $2\frac{5}{8}$ Miles

D. $1\frac{5}{8}$ Miles

9) $|8 - (45 \div |4 - 9|)| = ?$

A. 4

B. -1

C. 1

D. -4

10) The ratio of boys to girls in a school is 3:5. If there are 320 students in a school, how many boys are in the school?

A. 120   C. 60

B. 240   D. 160

11) When an integer is multiplied by itself, they can end in all of the following digits EXCEPT

A. 4, 7   C. 5, 6

B. 1, 9   D. 7, 8

12) The rectangle ABCD on the coordinate grid is translated 4 units down and 3 units to the left.

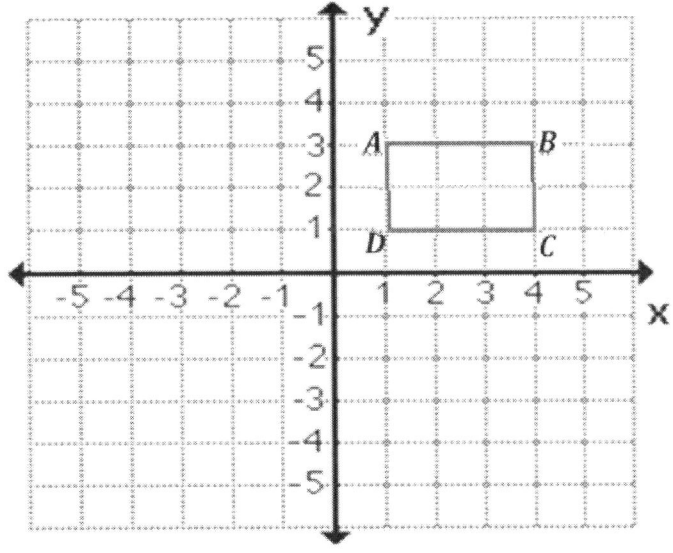

Which of the following describes this transformation?

A. $(x, y) \Rightarrow (x - 4, y - 3)$ C. $(x, y) \Rightarrow (x - 4, y + 3)$

B. $(x, y) \Rightarrow (x - 3, y - 4)$ D. $(x, y) \Rightarrow (x + 3, y - 4)$

13) Which graph shows a non-proportional linear relationship between x and y?

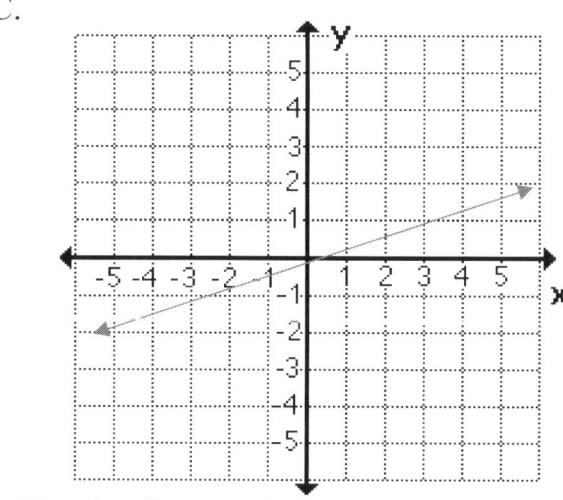

A.

B.

C.

D.

14) Use the diagram below to answer the question.

Given the lengths of the base and diagonal of the rectangle below, what is the

length of height h, in terms of s?

A. $4s\sqrt{6}$

B. $4s\sqrt{3}$

C. $14s$

D. $49s$

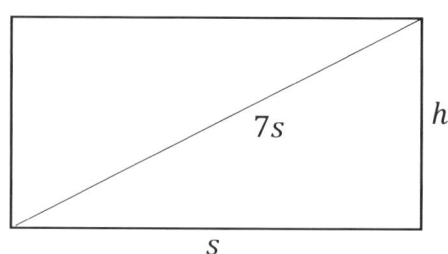

Use the chart below to answer the question.

Color	Number
White	15
Black	20
Beige	45

15) There are also purple marbles in the bag. Which of the following can **NOT** be the probability of randomly selecting a purple marble from the bag?

A. $\frac{1}{9}$

C. $\frac{4}{9}$

B. $\frac{3}{4}$

D. $\frac{5}{11}$

16) If the area of trapezoid is 120 cm, what is the perimeter of the trapezoid?

A. 24 cm

B. 36 cm

C. 48 cm

D. 96 cm

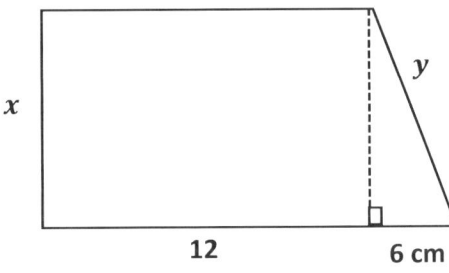

17) If a vehicle is driven 54 miles on Monday, 55 miles on Tuesday, and 47 miles on Wednesday, what is the average number of miles driven each day?

A. 52 Miles

C. 56 Miles

B. 57 Miles

D. 54 Miles

18) Find the area of a rectangle with a length of 124 feet and a width of 53 feet.

A. 6,752 sq. ft

C. 6,275 sq. ft

B. 6,572 sq. ft

D. 6,672 sq. ft

19) $76 \div \frac{1}{3} = ?$

A. 22.6

C. 168

B. 26

D. 228

20) With an 35% discount, Ella was able to save \$26.25 on a dress. What was the original price of the dress?

A. \$85

C. \$75

B. \$82

D. \$78

21) $\frac{9}{51}$ is equals to:

A. 1.254

C. 0.176

B. 1.056

D. 0.147

22) If 60% of A is 1,200, what is 15% of A?

A. 400

C. 300

B. 480

D. 250

23) If $(6.4 + 5.4 + 4.2) \times x = x$, then what is the value of x?

A. 0

C. -2

B. $\frac{3}{8}$

D. -6

24) Two dice are thrown simultaneously, what is the probability of getting a sum of 4 or 9?

A. $\frac{1}{12}$

C. $\frac{3}{10}$

B. $\frac{7}{36}$

D. $\frac{5}{35}$

25) Simplify $\dfrac{\frac{1}{3} - \frac{x+8}{6}}{\frac{x^2}{3} - \frac{5}{3}}$

A. $\dfrac{6-x}{x^2+10}$

C. $\dfrac{6-x}{x^2+10}$

B. $\dfrac{6+x}{2x^2+10}$

D. $\dfrac{-6-x}{2x^2-10}$

26) In the following figure, AB is the diameter of the circle. What is the circumference of the circle?

A. 13π

B. 14π

C. 11π

D. 24π

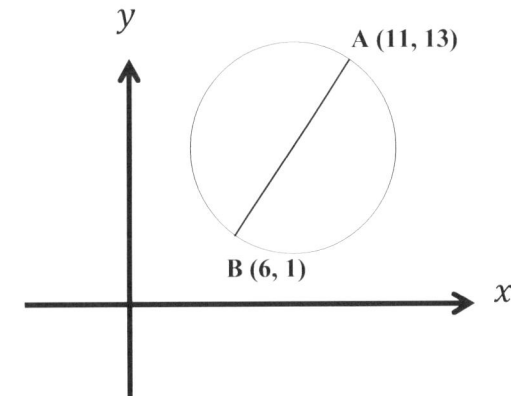

27) What is the value of x in the following equation?

$$4x^2 - 18 = 46$$

A. ± 5

C. ± 4

B. ± 8

D. ± 2

28) A circle has a diameter of 8 inches. What is its approximate area?

A. 56.25

C. 25.25

B. 50.24

D. 36.56

29) If 8 garbage trucks can collect the trash of 64 homes in a day. How many trucks are needed to collect in 160 houses?

A. 36

C. 16

B. 40

D. 20

Use the following table to answer question below.

DANIEL'S BIRD-WATCHING PROJECT	
Day	**Number of Raptors Seen**
Monday	?
Tuesday	7
Wednesday	18
Thursday	17
Friday	12
Mean	**13**

30) The above table shows the data Daniel collects while watching birds for one week. How many raptors did Daniel see on Monday?

A. 13

C. 11

B. 8

D. 18

31) $82.36 \div 0.09 =$?

A. 859.12

C. 925.22

B. 915.11

D. 9.211

32) A floppy disk shows 658,201 bytes free and 699,324 bytes used. If you delete a file of size 621,258 bytes and create a new file of size 596,657 bytes, how many free bytes will the floppy disk have?

A. 862,808

C. 608,502

B. 1,302,021

D. 682,802

33) 8 days 3 hours 20 minutes – 6 days 21 hours 35 minutes =?

A. 3 days 1 hours 55 minutes

B. 1 days 5 hours 45 minutes

C. 1 days 2 hours 55 minutes

D. 2 days 8 hours 15 minutes

34) The base of a right triangle is 14 feet, and the interior angles are 45-45-90. What is its area?

A. 98 square feet

C. 24 square feet

B. 48 square feet

D. 36 square feet

35) A circle is inscribed in a square, as shown below. The area of the circle is 25π cm². What is the area of the square?

A. 50 cm²

B. 20 cm²

C. 75 cm²

D. 100 cm²

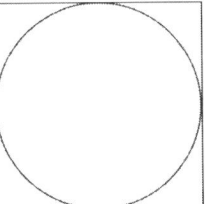

36) Increased by 80%, the number 15 becomes:

    A. 47

    B. 35

    C. 27

    D. 107

37) If $28 + x^{\frac{1}{3}} = 31$, then what is the value of $12 \times x$?

    A. 27   C. 423

    B. 54   D. 324

38) Triangle ABC is graphed on a coordinate grid with vertices at A $(-5, -2)$, B $(3, -9)$ and C $(4, 7)$. Triangle ABC is reflected over x axes to create triangle A'B'C'.

Which order pair represents the coordinate of C'?

    A. $(4, 7)$ C. $(-4, 7)$

    B. $(-7, -4)$ D. $(4, -7)$

39) Which set of ordered pairs represents y as a function of x?

    A. $\{(4, -2), (6, 0), (9, -4), (4, 5)\}$

    B. $\{(1, 6), (8, -2), (6, 2), (8, -3)\}$

    C. $\{(13, -8), (-3, 6), (8, 12), (-3, 7)\}$

    D. $\{(8, 2), (-3, 4), (1, 8), (5, -3)\}$

40) Which equation represents the statement "triple the difference between 6 times H and 4 gives 48".

A. $\dfrac{6H + 4}{3} = 48$

C. $3(6H - 4) = 48$

B. $6(2H + 4) = 48$

D. $3\dfrac{6H}{4} = 48$

41) David makes a weekly salary of \$400 plus 8% commission on his sales. What will his income be for a week in which he makes sales totaling \$1,100?

A. \$295

C. \$592

B. \$925

D. \$259

42) $9x^3y^4 + 5x^4y^5 - (7x^3y^4 - 5x^4y^5) = $ ___

A. $-2x^3y^4 - 8x^4y^5$

C. $10x^2y^3$

B. $x^4y^5 + 10x^3y^5$

D. $10x^4y^5 + 2x^3y^4$

43) The radius of circle A is four times the radius of circle B. If the circumference of circle A is 24π, what is the area of circle B?

A. 3π

C. 9π

B. 4π

D. 12π

44) If a box contains red and blue balls in ratio of 4: 5 red to blue, how many red balls are there if 60 blue balls are in the box?

A. 12

C. 24

B. 48

D. 96

45) The width of a box is one fifth of its length. The height of the box is one sixth of its width. If the length of the box is 60 cm, what is the volume of the box?

A. 42 cm^3

C. 1,440 cm^3

B. 440 cm^3

D. 2,880 cm^3

46) A square measures 9 inches on one side. By how much will the area be decreased if its length is increased by 7 inches and its width decreased by 5 inches.

A. 17 sq. decreased

C. 19 sq. decreased

B. 14 sq. decreased

D. 15 sq. decreased

47) How many 4 × 4 squares can fit inside a rectangle with a height of 44 and width of 12?

A. 66

C. 99

B. 33

D. 88

STOP

IF YOU FINISH BEFORE TIME IS CALLED, YOU MAY CHECK YOUR WORK ON THIS SECTION ONLY. DO NOT TURN TO ANY OTHER SECTION IN THE TEST.

ISEE Upper Level Practice Test 2

Mathematics

Quantitative Reasoning

- ❖ **37 Questions.**
- ❖ **Total time for this test: 35 Minutes**.
- ❖ **Calculators are not allowed at the test.**

Administered *Month Year*

1) The area of a circle is less than 100π. Which of the following can be the circumference of the circle?

    A. $22\,\pi$ C. $34\,\pi$

    B. $18\,\pi$ D. $42\,\pi$

2) The circle graph below shows all Mr. Green's expenses for last month. If he spent $360 on his car, how much did he spend for his rent?

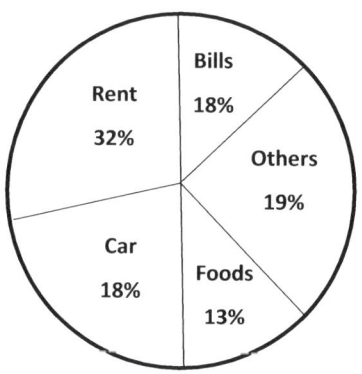

Mr. Green's monthly expenses

    A. $600

    B. $740

    C. $640

    D. $540

3) Etta drives from her house to work at an average speed of 64 miles per hour and she drives at an average speed of 60 miles per hour when she was returning home. What was her minimum speed on the round trip in miles per hour?

    A. 60   C. 64

    B. 62   D. Cannot be determined

4) How much greater is the value of $4x + 9$ than the value of $4x - 5$?

    A. 6   C. 45

    B. 14   D. 18

5) Oscar purchased a new hat that was on sale for $8.98. The original price was $18.32. What percentage discount was the sale price?

A. 34.90%   C. 50.98%

B. 50.11%   D. 93.40%

6) If $f(x) = 2x^2 + 12$, what is the smallest possible value of $f(x)$?

A. 14   C. 12

B. 6   D. 8

7) If the sum of the positive integers from 1 to $n + 1$ is 3,875, and the sum of the positive integers from $n + 2$ to $2n$ is 9,328, which of the following represents the sum of the positive integers from 1 to $2n$ inclusive?

A. 12,903   C. 9,680

B. 7,302   D. 13,203

8) Triangle ABC is similar to triangle ADE. What is the length of side EC?

A. 6

B. 18

C. 12

D. 24

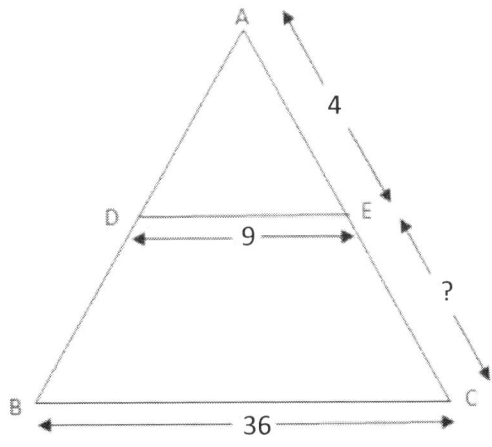

9) Which of the following statements is correct, according to the graph below?

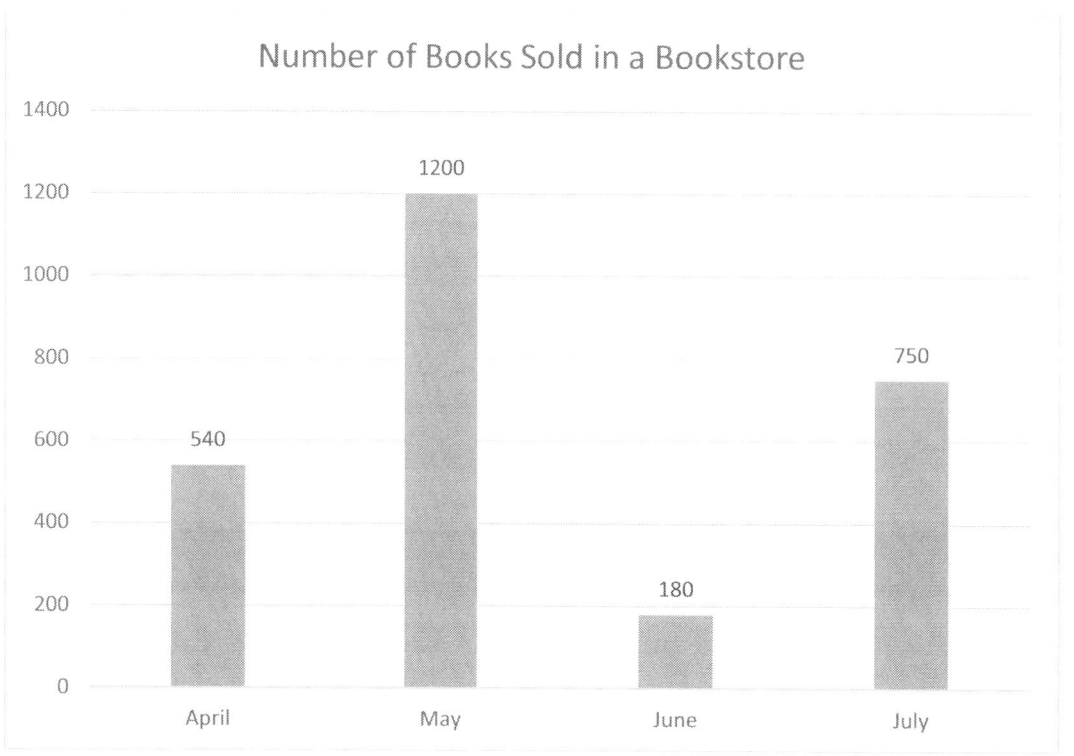

A. Number of books sold in June was greater than one third the number of books sold in July.

B. Number of books sold in July was One third the number of books sold in May.

C. Number of books sold in April was triple the number of books sold in June.

D. Number of books sold in July was equal to the number of books sold in April plus the number of books sold in June.

10) What is the prime factorization of 180?

A. $2 \times 3 \times 3 \times 3$

C. 3×7

B. $2 \times 2 \times 5 \times 7$

D. $2 \times 2 \times 3 \times 3 \times 5$

11) List A consists of the numbers {3, 4, 8, 12, 21}, and list B consists of the numbers {1, 7, 10, 15, 19}.

If the two lists are combined, what is the median of the combined list?

A. 9   C. 8

B. 14   D. 18

12) If 12 inches on a map represents an actual distance of 80 feet, then what actual distance does 36 inches on the map represent?

A. 24   C. 84

B. 400   D. 240

13) If Jim adds 60 stamps to his current stamp collection, the total number of stamps will be equal to $\frac{8}{7}$ the current number of stamps. If Jim adds 20% more stamps to the current collection, how many stamps will be in the collection?

A. 420   C. 405

B. 840   D. 504

14) A gas tank can hold 21 gallons when it is $\frac{3}{7}$ full. How many gallons does it contain when it is full?

A. 12   C. 10

B. 9   D. 17.5

15) A basket contains 24 balls and the average weight of each of these balls is 15 g. The six heaviest balls have an average weight of 18 g each. If we remove the six heaviest balls from the basket, what is the average weight of the remaining balls?

A. 18

C. 8.5

B. 14

D. 9.5

16) A bag contains 18 balls: three green, seven black, four blue, two brown, two red and one white. If 12 balls are removed from the bag at random, what is the probability that a brown ball has been removed?

A. $\frac{1}{12}$

C. $\frac{2}{3}$

B. $\frac{1}{18}$

D. $\frac{1}{9}$

17) If $2x + y = 8$ and $2x - y = 4$ then what is the value of $(4x^2 - y^2)$?

A. 32

C. 16

B. 20

D. 80

18) What's The ratio of boys and girls in a class is 4:7. If there are 77 students in the class, how many more boys should be enrolled to make the ratio 1:1?

A. 49

C. 28

B. 8

D. 21

19) The area of rectangle ABCD is 256 square inches. If the length of the rectangle is four times the width, what is the perimeter of rectangle ABCD?

A. 80   C. 45

B. 20   D. 60

20) The sum of 7 numbers is greater than 350 and less than 420 Which of the following could be the average (arithmetic mean) of the numbers?

A. 6   C. 75

B. 50   D. 55

21) Which of the following expressions gives the value of z in terms of a, b, and c from the following equation?

$$a = [\frac{2cz}{3b}]^2$$

A. $z = 6ac^2b^2$ C. $z = \frac{cb}{\sqrt{2a}}$

B. $z = \frac{3b\sqrt{a}}{2C}$ D. $z = [\frac{3b}{2ac}]^2$

Quantitative Comparisons

Direction: Questions 22 to 37 are Quantitative Comparisons Questions. Using the information provided in each question, compare the quantity in column A to the quantity in Column B. Choose on your answer sheet grid

- A if the quantity in Column A is greater
- B if the quantity in Column B is greater
- C if the two quantities are equal
- D if the relationship cannot be determined from the information given

22)

Column A	**Column B**
The average of 14, 28, and 18	18

23)

Column A	**Column B**
$14 \times 268 \times 15$	$17 \times 268 \times 15$

24) x is an integer

Column A	**Column B**
x	$\dfrac{x}{-7}$

25)

Column A	**Column B**
The greatest value of x in	The greatest value of x in
$4\|3x - 8\| = 4$	$4\|3x + 8\| = 4$

26) x is an integer

Column A	Column B
$(x)^4(x)^2$	$(x^4)^2$

27)

Column A	Column B
2^8	$\sqrt[3]{512}$

28)

Column A	Column B
6	$(24)^{\frac{1}{2}}$

29) The selling price of a sport jacket including 40% discount is $180.

Column A	Column B
Original price of the sport jacket	$275

30)

Column A	Column B
$(0.55)^{22}$	$(0.55)^{21}$

31)

Column A	Column B
The probability that event x will occur.	The probability that event x will not occur.

32)

Column A	**Column B**
The probability of rolling a 2 on a die and getting heads on a coin toss.	The probability of rolling an even number on a die and picking a spade from a deck of 52 cards.

33)

Column A	**Column B**
$(\frac{1}{5})^3$	5^{-3}

34) $3x + 3 > 6x$

Column A	**Column B**
x	-2

35)

Column A	**Column B**
$1.65	Sum of five quarter, three nickels, and four pennies

36) x is an even integer, and y is an odd integer, in a certain game an even number is considered greater than an odd number?

Column A	**Column B**
$(y)(x + y)$	$(x - y)^2 + y$

37) $x^2 - 2x - 16 = 32$

<u>Column A</u>	<u>Column B</u>
x	1

STOP

YOU FINISH BEFORE TIME IS CALLED; YOU MAY CHECK YOUR WORK ON THIS SECTION ONLY. DO NOT TURN TO ANY OTHER SECTION IN THE TEST.

ISEE Upper Level Practice Test 2

Mathematics

Mathematics Achievement

- ❖ **47 Questions.**
- ❖ **Total time for this test: 40 Minutes.**
- ❖ **Calculators are not allowed at the test.**

Administered *Month Year*

1) $\frac{13}{26}$ is equal to:

    A. 5.0   C. 0.05

    B. 0.50   D. 0.005

2) 10 less than triple a positive integer is 29. What is the integer?

    A. 27   C. 13

    B. 16   D. 33

3) Which of the following points lies on the line $3x + 4y = 11$?

    A. $(3, -1)$ C. $(-2, 0)$

    B. $(-3, 1)$ D. $(2, 1)$

4) An angle is equal to one eighth of its supplement. What is the measure of that angle?

    A. 15   C. 20

    B. 22.5   D. 18

5) 3.2 is what percent of 40?

    A. 1.6   C. 18

    B. 8   D. 6

6) Right triangle ABC has two legs of lengths 10 cm (AB) and 24 cm (AC). What is the length of the third side (BC)?

    A. 14 cm   C. 13 cm

    B. 12 cm   D. 26 cm

7) $\frac{1}{4b^2} + \frac{1}{4b} = \frac{1}{2b^2}$, then $b =$?

A. $-\frac{2}{8}$

C. $-\frac{2}{10}$

B. 1

D. 4

8) If $\frac{|7+x|}{3} \leq 6$, then which of the following is correct?

A. $-11 \leq x \leq 25$

C. $-11 \leq x \leq 11$

B. $-25 \leq x \leq 11$

D. $-25 \leq x \leq 25$

9) The cost, in thousands of dollars, of producing x thousands of textbooks is

$C(x) = x^2 + 2x$. The revenue, also in thousands of dollars, is $R(x) = 15x$. find

the profit or loss if 8 textbooks are produced. (profit = revenue – cost)

A. $20 profit

C. $120 loss

B. $40 profit

D. $80 loss

10) Ella (E) is 5 years older than her friend Ava (A) who is 4 years younger than

her sister Sofia (S). If E, A and S denote their ages, which one of the following

represents the given information?

A. $\begin{cases} E = A + 5 \\ S = A + 4 \end{cases}$

C. $\begin{cases} A = E + 5 \\ A = S + 4 \end{cases}$

B. $\begin{cases} E = A - 5 \\ S = A - 4 \end{cases}$

D. $\begin{cases} A = E - 5 \\ A = S - 4 \end{cases}$

11) Simplify $9x^3y^2(2x^2y)^3 =$

A. $36x^5y^4$ C. $18x^9y^5$

B. $72x^5y^9$ D. $72x^9y^5$

12) Which is the longest time?

A. 28 hours   C. 1 days and 2 hours

B. 4,210 minutes   D. 58,800 seconds

13) Write 467 in expanded form, using exponents.

A. $(4 \times 10^3) + (6 \times 10^2) + (7 \times 10)$

B. $(4 \times 10^2) + (6 \times 10^1) - 10$

C. $(4 \times 10^2) + (6 \times 10^1) + 7$

D. $(4 \times 10^1) + (6 \times 10^2) + 7$

14) A company pays its writer $4 for every 400 words written. How much will a writer earn for an article with 540 words?

A. $4.5   C. $5.4

B. $ 2.8   D. $54

15) A circular logo is enlarged to fit the lid of a jar. The new diameter is 70% larger than the original. By what percentage has the area of the logo increased?

A. 30%   C. 36%

B. 69%   D. 6%

16) $84.24 \div 0.08 = ?$

    A. 10.53   C. 10.053

    B. 1,053   D. 105.3

17) A bread recipe calls for $4\frac{2}{3}$ cups of flour. If you only have $3\frac{5}{6}$ cups, how much more flour is needed?

    A. 6   C. 5

    B. $\frac{1}{6}$ D. $\frac{5}{6}$

18) The equation of a line is given as: $y = 7x - 5$. Which of the following points does not lie on the line?

    A. $(0, -5)$ C. $(-3, 8)$

    B. $(-1, -12)$ D. $(1, 2)$

19) A circle has a diameter of 10 inches. What is its approximate circumference?

    A. 21   C. 15

    B. 31   D. 45

20) What's the area of the non-shaded part of the following figure?

    A. $116 \; cm^2$

    B. $136 \; cm^2$

    C. $88 \; cm^2$

    D. $224 \; cm^2$

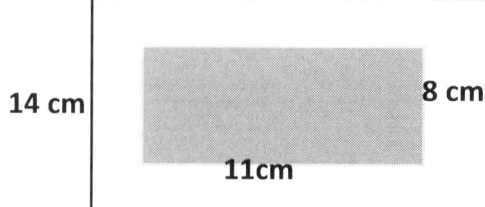

21) What is the area of an isosceles right triangle with hypotenuse that measures 10 cm?

A. 25 cm

C. $5\sqrt{2}$ cm

B. 10 cm

D. 18 cm

22) The drivers at G & G trucking must report the mileage on their trucks each week. The mileage reading of Ed's vehicle was 42,857 at the beginning of one week, and 43,128 at the end of the same week. What was the total number of miles driven by Ed that week?

A. 27 Miles

C. 271 Miles

B. 127 Miles

D. 1,721 Miles

23) What is the maximum value for y if $y = -(x-3)^2 + 8$?

A. -8

C. 3

B. -3

D. 8

24) What is the solution of the following system of equations?

$$\begin{cases} -3x + y = -5 \\ 4x + 3y = 11 \end{cases}$$

A. $(-1, 3)$

C. $(1, 3)$

B. $(2, 1)$

D. $(2, -1)$

25) What is the area of an isosceles right triangle that has one leg that measures 4 cm?

C. $8\sqrt{2}$ cm^2

A. 8 cm^2

D. 24 cm^2

B. 16 cm^2

26) Which of the following is a factor of both $x^2 + 4x - 45$ and $x^2 - 12x + 35$?

A. $(x - 5)$

B. $(x + 5)$

C. $(x - 7)$

D. $(x + 7)$

27) If $x + y = 18$, what is the value of $2x + 2y$?

A. 9

B. 18

C. 36

D. 42

28) A car uses 14 gallons of gas to travel 350 miles. How many miles per gallon does the car use?

A. 18 miles per gallon

B. 26 miles per gallon

C. 25 miles per gallon

D. 20 miles per gallon

29) $(5x + 5)(x + 4) =$

A. $5x + 20$

B. $5x^2 + 20x + 24$

C. $5x^2 + 25x + 20$

D. $5x^2 + 5$

30) If $x \blacksquare y = \sqrt{x^2 + 4y}$, what is the value of $3 \blacksquare 4$?

A. $\sqrt{35}$

B. 5

C. 12

D. 4

31) There are two equal tanks of water. If $\frac{4}{7}$ of a tank contains 210 liters of water, what is the capacity of the three tanks of water together?

A. 120

B. 420

C. 90

D. 240

32) The average weight of 16 girls in a class is 45 kg and the average weight of 24 boys in the same class is 55 kg. What is the average weight of all the 46 students in that class?

A. 48

B. 51

C. 50.1

D. 49.1

33) If x is 35% percent of 740, what is x?

A. 225

B. 252

C. 552

D. 525

34) $\begin{array}{r} 28\ \text{hr.}\ \ 15\ \text{min.} \\ -\ 19\ \text{hr.}\ \ 37\ \text{min.} \\ \hline \end{array}$

A. $8\ hr.\ 22\ min.$

B. $8\ hr.\ 38\ min.$

C. $13\ hr.\ 22\ min.$

D. $13\ hr.\ 38\ min.$

35) What is the number of cubic feet of soil needed for a flower box 5 feet long, 4 inches wide, and 6 feet deep?

A. 10 cubic feet

B. 120 cubic feet

C. $\frac{20}{6}$ cubic feet

D. 9 cubic feet

36) Karen is 9 years older than her sister Michelle, and Michelle is 3 years younger than her brother David. If the sum of their ages is 84, how old is Michelle?

A. 27

B. 24

C. 22

D. 25

37) Mario loaned Jett $950 at a yearly interest rate of 2%. After two years what is the interest owned on this loan?

A. $1,320

C. $20

B. $125

D. $38

38) Mason just got hired for on-the-road sales and will travel about 2,100 miles a week during a 42-hour work week. If the time spent traveling is $\frac{2}{7}$ of his week, how many hours a week will he be on the road?

A. Mason spends about 7 hours of his 42-hour work week on the road.

B. Mason spends about 10 hours of his 42-hour work week on the road.

C. Mason spends about 21 hours of his 42-hour work week on the road.

D. Mason spends about 12 hours of his 42-hour work week on the road.

39) A shirt costing $570 is discounted 30%. After a month, the shirt is discounted another 18%. Which of the following expressions can be used to find the selling price of the shirt?

A. $(570)(0.7)$

C. $(570)(0.48) - (570)(0.30)$

B. $(570) - 570(0.70)$

D. $(570)(0.7)(0.82)$

40) What is the reciprocal of $\frac{x^4}{18}$?

A. $\frac{18}{x^4} - 2$

C. $\frac{18}{x^4} + 2$

B. $\frac{54}{x^4}$

D. $\frac{18}{x^4}$

41) Given that $x = 0.3$ and $y = 4$ what is the value of $5x^2(y + 6)$?

   A. 18   C. 3.6

   B. 1.8   D. 36

42) Calculate the area of a parallelogram with a base of 4 feet and height of 5.6 feet.

   A. 12.4 square feet   C. 22.4 square feet

   B. 14.4 square feet   D. 26.4 square feet

43) In a school, the ratio of number of boys to girls is 7:5. If the number of boys is 280, what is the total number of students in the school?

   A. 120   C. 240

   B. 480   D. 1,420

44) A tree 32 feet tall casts a shadow 14 feet long. Jack is 8 feet tall. How long is Jack's shadow?

   A. 3.5 ft C. 35 ft

   B. 4.6 ft D. 46 ft

45) What is the area of the shaded region if the diameter of the bigger circle is 16 inches and the diameter of the smaller circle is 6 inches?

   A. 20 π inch^2

   B. 55 π inch^2

   C. 35 π inch^2

   D. 124 π inch^2

46) What is the result of the expression?

$$\begin{vmatrix} 3 & 5 \\ -2 & -2 \\ -8 & 6 \end{vmatrix} + \begin{vmatrix} 1 & -1 \\ 5 & -1 \\ 7 & -4 \end{vmatrix} =$$

A. $\begin{vmatrix} 4 & -4 \\ 3 & 1 \\ 1 & 2 \end{vmatrix}$

C. $\begin{vmatrix} 4 & 4 \\ 3 & -3 \\ -1 & 2 \end{vmatrix}$

B. $\begin{vmatrix} 2 & -1 \\ -3 & -2 \\ -10 & -4 \end{vmatrix}$

D. $\begin{vmatrix} 2 & -1 \\ -5 & 3 \\ -12 & -4 \end{vmatrix}$

47) How many square feet of tile is needed for 7 feet by 7 feet room?

A. 14 square feet

C. 49 square feet

B. 28 square feet

D. 121 square fee

STOP

IF YOU FINISH BEFORE TIME IS CALLED, YOU MAY CHECK YOUR WORK ON THIS SECTION ONLY. DO NOT TURN TO ANY OTHER SECTION IN THE TEST.

Answers and Explanations

ISEE Upper Level Practice Tests

Answer Key

❋ Now, it's time to review your results to see where you went wrong and what areas you need to improve!

ISEE Upper Level Practice Test 1 - Mathematics

Quantitative Reasoning						Mathematics Achievement								
1	D	16	C	31	D	1	A	16	C	31	B	46	A
2	C	17	C	32	A	2	B	17	A	32	D	47	B
3	B	18	D	33	D	3	C	18	B	33	B		
4	C	19	A	34	D	4	B	19	D	34	A		
5	C	20	D	35	A	5	C	20	C	35	D		
6	C	21	A	36	B	6	D	21	C	36	C		
7	B	22	B	37	A	7	B	22	C	37	D		
8	D	23	B			8	D	23	A	38	D		
9	B	24	B			9	C	24	B	39	D		
10	B	25	C			10	A	25	D	40	C		
11	A	26	B			11	D	26	A	41	C		
12	D	27	C			12	B	27	C	42	D		
13	A	28	A			13	B	28	B	43	C		
14	A	29	A			14	B	29	D	44	B		
15	A	30	A			15	D	30	C	45	C		

Answers and Explanations
ISEE Upper Level Practice Tests

ISEE Upper Level Practice Test 2 - Mathematics

<table>
<tr><td colspan="6">Quantitative Reasoning</td></tr>
<tr><td>1</td><td>B</td><td>16</td><td>D</td><td>31</td><td>D</td></tr>
<tr><td>2</td><td>C</td><td>17</td><td>A</td><td>32</td><td>B</td></tr>
<tr><td>3</td><td>D</td><td>18</td><td>D</td><td>33</td><td>C</td></tr>
<tr><td>4</td><td>B</td><td>19</td><td>A</td><td>34</td><td>D</td></tr>
<tr><td>5</td><td>C</td><td>20</td><td>D</td><td>35</td><td>A</td></tr>
<tr><td>6</td><td>C</td><td>21</td><td>B</td><td>36</td><td>A</td></tr>
<tr><td>7</td><td>D</td><td>22</td><td>A</td><td>37</td><td>D</td></tr>
<tr><td>8</td><td>C</td><td>23</td><td>B</td><td></td><td></td></tr>
<tr><td>9</td><td>C</td><td>24</td><td>D</td><td></td><td></td></tr>
<tr><td>10</td><td>D</td><td>25</td><td>A</td><td></td><td></td></tr>
<tr><td>11</td><td>A</td><td>26</td><td>D</td><td></td><td></td></tr>
<tr><td>12</td><td>D</td><td>27</td><td>C</td><td></td><td></td></tr>
<tr><td>13</td><td>D</td><td>28</td><td>A</td><td></td><td></td></tr>
<tr><td>14</td><td>B</td><td>29</td><td>A</td><td></td><td></td></tr>
<tr><td>15</td><td>B</td><td>30</td><td>B</td><td></td><td></td></tr>
</table>

<table>
<tr><td colspan="8">Mathematics Achievement</td></tr>
<tr><td>1</td><td>B</td><td>16</td><td>B</td><td>31</td><td>D</td><td>46</td><td>C</td></tr>
<tr><td>2</td><td>C</td><td>17</td><td>D</td><td>32</td><td>B</td><td>47</td><td>C</td></tr>
<tr><td>3</td><td>D</td><td>18</td><td>C</td><td>33</td><td>B</td><td></td><td></td></tr>
<tr><td>4</td><td>C</td><td>19</td><td>B</td><td>34</td><td>B</td><td></td><td></td></tr>
<tr><td>5</td><td>B</td><td>20</td><td>B</td><td>35</td><td>A</td><td></td><td></td></tr>
<tr><td>6</td><td>D</td><td>21</td><td>A</td><td>36</td><td>B</td><td></td><td></td></tr>
<tr><td>7</td><td>B</td><td>22</td><td>C</td><td>37</td><td>D</td><td></td><td></td></tr>
<tr><td>8</td><td>B</td><td>23</td><td>D</td><td>38</td><td>D</td><td></td><td></td></tr>
<tr><td>9</td><td>B</td><td>24</td><td>B</td><td>39</td><td>D</td><td></td><td></td></tr>
<tr><td>10</td><td>A</td><td>25</td><td>A</td><td>40</td><td>D</td><td></td><td></td></tr>
<tr><td>11</td><td>D</td><td>26</td><td>A</td><td>41</td><td>B</td><td></td><td></td></tr>
<tr><td>12</td><td>B</td><td>27</td><td>C</td><td>42</td><td>C</td><td></td><td></td></tr>
<tr><td>13</td><td>C</td><td>28</td><td>C</td><td>43</td><td>B</td><td></td><td></td></tr>
<tr><td>14</td><td>C</td><td>29</td><td>C</td><td>44</td><td>A</td><td></td><td></td></tr>
<tr><td>15</td><td>B</td><td>30</td><td>B</td><td>45</td><td>B</td><td></td><td></td></tr>
</table>

Score Your Test

ISEE Upper Level scores are broken down by four sections: Verbal Reasoning, Reading Comprehension, Quantitative Reasoning, and Mathematics Achievement. A sum of the ALL sections is also reported. The Essay section is scored separately. For the Upper Level ISEE, the score range is 760 to 940, the lowest possible score a student can earn is 760 and the highest score is 940 for each section. A student receives 1 point for every correct answer. There is no penalty for wrong or skipped questions.

The total scaled score for an Upper Level ISEE test is the sum of the scores for all sections. A student will also receive a percentile score of between 1-99% that compares that student's test scores with those of other test takers of same grade and gender from the past 3 years. When a student receives her/his score, the percentile score is also be broken down into a stanine and the stanines are ranging from 1–9. Most schools accept students with scores of 5–9. The ideal candidate has scores of 6 or higher.

The following charts provide an estimate of students ISEE percentile rankings for the practice tests, compared against other students taking these tests. Keep in mind that these percentiles are estimates only, and your actual ISEE percentile will depend on the specific group of students taking the exam in your year.

Percentile Rank	Stanine
1 – 3	1
4 – 10	2
11- 22	3
23 – 39	4
40 – 59	5
60 – 76	6
77- 88	7
89 – 95	8
96 – 99	9

ISEE Upper Level Quantitative Reasoning Percentiles

Grade Applying to	25th Percentile	50th Percentile	75th Percentile
9th	850	880	895
10th	855	885	900
11th	860	890	905
12th	864	892	908

Use the next table to convert ISEE Upper level raw score to scaled score for application to 9th - 12th grade.

ISEE Upper Level Scaled Scores

Raw Score	Quantitative Reasoning Report Range		Mathematics Achievement Report Range		Raw Score	Quantitative Reasoning Report Range		Mathematics Achievement Report Range	
0	760	760	760	760	26	900	885	885	865
1	770	765	770	765	27	905	890	885	865
2	780	770	780	770	28	910	895	890	870
3	790	775	790	775	29	910	900	890	870
4	800	780	800	780	30	915	905	895	875
5	810	785	810	785	31	920	910	895	875
6	820	790	820	790	32	925	915	900	880
7	825	795	825	795	33	930	920	900	880
8	830	800	830	800	34	930	925	905	885
9	835	805	835	805	35	935	930	905	885
10	840	810	840	810	36	935	935	910	890
11	845	815	845	815	37	940	940	910	890
12	850	820	850	820	38			915	895
13	855	825	855	825	39			920	900
14	860	830	855	830	40			925	905
15	865	835	860	835	41			925	910
16	870	840	860	840	42			930	915
17	875	845	865	840	43			930	920
18	880	845	865	845	44			935	925
19	880	850	870	845	45			935	930
20	885	855	870	850	46			940	935
21	885	860	875	850	47			940	940
22	890	865	875	855					
23	890	870	875	855					
24	895	875	880	860					
25	895	880	880	860					

Answers and Explanations

ISEE - Upper Level

Practice Tests 1: Quantitative Reasoning

1) Answer: D.

A factor must divide evenly into its multiple. 14 cannot be a factor of 80 because 80 divided by 14 = 5.71

2) Answer: C.

$$\text{Probability} = \frac{number\ of\ desired\ outcomes}{number\ of\ total\ outcomes}$$

In this case, a desired outcome is selecting either a blue or a yellow marble. Combine the number of blue and yellow marbles: $9 + 6 = 15$ and divide this by the total number of marbles: $9 + 6 + 10 = 25$. The probability is $\frac{15}{25} = \frac{3}{5}$.

3) Answer: B.

The distance on the map is proportional to the actual distance between the two cities. Use the information to set up a proportion and then solve for the unknown number of actual miles: $\dfrac{6\ miles}{\frac{1}{4}\ inches} = \dfrac{x\ miles}{18\ inches}$

Cross multiply and simplify to solve for the x:

$$\frac{6 \times 18}{\frac{1}{4}} = x\ miles \quad \rightarrow \quad \frac{108}{\frac{1}{4}} = 108 \times 4 = 432\ miles$$

4) Answer: C.

The diagonal of the square is 12. Let x be the side.

Use Pythagorean Theorem: $a^2 + b^2 = c^2$

$x^2 + x^2 = 12^2 \Rightarrow 2x^2 = 12^2 \Rightarrow 2x^2 = 144 \Rightarrow x^2 = 72 \Rightarrow x = \sqrt{72}$

The area of the square is: $\sqrt{72} \times \sqrt{72} = 72$

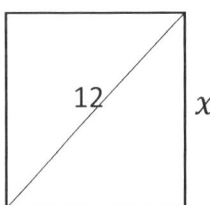

5) Answer: C.

To find what percent A is of B, divide A by B, then multiply that number by 100%:

$23.75 \div 58.25 = 0.4077 \times 100\% = 40.77\%$

This is approximately 41 %.

6) Answer: C.

The total cost of the phone call can be represented by the equation: TC = $6.00 + 0.3(x - 3)$, where x is the duration of the call after the first three minutes. In this case, $x = 20$. Substitute the known values into the equation and solve:

TC = $6.00 + $0.3 \times (20 - 3)$

TC = $6.00 + $5.10 \Rightarrow$ TC = $11.10

7) Answer: B.

Let b be the amount of time Alec can do the job, then,

$$\frac{1}{a} + \frac{1}{b} = \frac{1}{90} \rightarrow \frac{1}{180} + \frac{1}{b} = \frac{1}{90} \rightarrow \frac{1}{b} = \frac{1}{90} - \frac{1}{180} = \frac{1}{180}$$

Then: $b = 180$ minutes

8) Answer: D.

Begin by calculating Elise total earnings after 24 hours:

24 hours $\times$ $6.50 per hour = $156

Next, divide this total by Bob's hourly rate to find the number of hours Bob would need to work: $156 ÷ $6.00 per hour = 26 hours

9) Answer: B.

The length of MN is equal to: $4x + 8x = 12x$

Then: $12x = 48 \rightarrow x = \frac{48}{12} = 4$

The length of ON is equal to: $8x = 8 \times 4 = 32$ cm

10) Answer: B.

Number of Physics book: $0.3 \times 400 = 120$

Number of Chemistry books: $0.09 \times 400 = 36$

Product of number of Physics and number of Chemistry books: $120 \times 36 = 4,320$

11) Answer: A.

First, find the number.

Let x be the number. Write the equation and solve for x.

160 % of a number is 48, then: $1.6 \times x = 48 \rightarrow x = 48 ÷ 1.6 = 30$

40% of 30 is: $0.40 \times 30 = 12$

12) **Answer: D.**

$5y + 6 < 41 \rightarrow 5y < 41 - 6 \rightarrow 5y < 35 \rightarrow y < 7$

Only choice D (6.5) is less than 7.

13) **Answer: A.**

Begin by examining the sequence to find the pattern. The difference between 3 and 6 is 3; moving from 6 to 11 requires 5 to be added; moving from 11 to 18 requires 7 to be added; moving from 18 to 27 requires 9 to be added. The pattern emerges here — adding by consecutive odd integers. The 6th term is equal to $27 + 11 = 39$, and the 7th term is equal to $39 + 13 = 52$.

14) **Answer: A.**

Write equations based on the information provided in the question: Emily = Daniel,

Emily = 8 Claire,   Daniel = 21 + Claire

$\rightarrow$8 Claire = 21 + Claire $\rightarrow$8 Claire – Claire = 21

7 Claire = 21,   Claire = 3

15) **Answer: A.**

The general slope-intercept form of the equation of a line is $y = mx + b$, where m is the slope and b is the y-intercept.

By substitution of the given point and given slope: $-2 = (2)(5) + b$

So, $b = -2 - 10 = -12$, and the required equation is $y = 2x - 12$

16) **Answer: C.**

Let's choose \$100 for the sales of the supermarket. If the sales increases by 30 percent in April, the final amount of sales at the end of April will be \$100 + (30%) × (\$100) = \$130. If sales then decreased by 30 percent in May, the final amount of sales at the end of May will be \$130 – (\$130) × (30%) = 130 – 39 = \$91

The final sales of \$91 is 91% of the original price of \$100.

Therefore, the sales decreased by 9% overall.

17) **Answer: C.**

If each book weighs $\frac{1}{8}$ pound, then 1 pound = 8 books. To find the number of books in

12 pounds, simply multiply this 8 by 12: $12 \times 8 = 96$

18) Answer: D.

Multiplying each side of $-3x + y = 5$ by -2 gives $6x - 2y = -10$.

Adding each side of $6x - 2y = -10$ to the corresponding side of $-6x + 4y = 14$ gives $2y = 4$ or $y = 2$.

Finally, substituting 2 for y in $-6x + 4y = 14$ gives $-6x + 4(2) = 14$ or $x = -1$.

19) Answer: A.

The time it takes to drive from city A to city B is: $\frac{4,800}{76} = 63.16$

20) Answer: D.

Let x be the number. Write the equation and solve for x.

$\frac{3}{7} \times 28 = \frac{4}{9}x \rightarrow \frac{3 \times 28}{7} = \frac{4x}{9}$, use cross multiplication to solve for x.

$27 \times 28 = 4x \times 7 \Rightarrow 756 = 28x \Rightarrow x = 27$

21) Answer: A.

The question is this: 496.40 is what percent of 584?

Use percent formula:

Part $= \frac{percent}{100} \times$ whole

$496.40 = \frac{percent}{100} \times 584 \rightarrow 496.40 = \frac{percent \times 584}{100} \rightarrow 49,640 = $ percent $\times 584$

Then, Percent $= \frac{49,640}{584} = 85$

496.40 is 85% of 584. Therefore, the discount is: $100\% - 85\% = 15\%$

22) Answer: B.

Recall that the first and only even prime number is 2. The other prime numbers are: 3, 5, 7, 11, 13, 17, etc. They are all odd numbers less than 10 are 3, 5, 7, so the sum of members in Set A is $3 + 5 + 7 = 15$. So, the correct answer is B.

23) Answer: B.

Column A: $1.8\% = 0.018$

Column B: $\frac{1}{6} = 0.166$

0.166 is greater than 0. 018

24) Answer: B.

Column A: $\dfrac{11 + 14 + 19 + 21 + 34}{5} = \dfrac{99}{5} = 19.8$

Column B: $\dfrac{22 + 24 + 34 + 28}{4} = \dfrac{108}{4} = 27$

25) Answer: C.

Column A: $8 + 12(9 - 8) = 20$

Column B: $12 + 8(9 - 8) = 20$

26) Answer: B.

The posts are placed 11.5 feet apart. Since a post is needed at the very beginning as well as at the end, B requires 15 posts. $138 \div 11.5 = 12 \implies 12 + 3 = 15$

27) Answer: C.

First find the value of x.

$\dfrac{x}{56} = \dfrac{3}{14} \rightarrow 14x = 3 \times 56 = 168 \rightarrow x = \dfrac{168}{14} = 12$

Column A: $\dfrac{6}{x} = \dfrac{6}{12} = \dfrac{1}{2}$

28) Answer: A.

First convert hours to minutes. 1 hours 45 minutes $= 1 \times 60 + 45 = 105$ minutes.

Machine D makes b rolls of steel in 15 minutes. So, it makes 7 sets of b in 100 minutes. $105 \div 15 = 7$ sets of b.

Machine E operates for 1 hours, making b rolls per 12 minutes. So, it makes a total of $5b$ rolls. Therefore, machine D makes more rolls, and Column A is greater.

29) Answer: A.

The ratio of boys to girls in a class is 10 to 12. Therefore, ratio of girls to the entire class is 12 out of 22. $\dfrac{12}{22} = \dfrac{6}{11} > \dfrac{1}{5}$

30) Answer: A.

Let us calculate each probability individually:

That probability that the first marble is blue $= \dfrac{8}{14} = \dfrac{4}{7}$

The probability that the second marble is blue $= \dfrac{7}{13}$

Column A: The probability that both marbles are blue: $\frac{4}{7} \times \frac{7}{13} = \frac{28}{91} = \frac{4}{13}$

The probability that the marble is green $= \frac{6}{14} = \frac{3}{7}$

The probability that the second marble is blue $= \frac{5}{13}$

Column B: The probability that the first marble is green, but the second is blue

$= \frac{3}{7} \times \frac{5}{13} = \frac{15}{91}$

Column A is greater. $\frac{4}{13} > \frac{15}{91}$.

31) Answer: D.

First, solve the expression for x. $\frac{x}{9} = y^2 \rightarrow x = 9y^2$

Plug in different values for y and find the values of x.

Let's choose $y = 0 \rightarrow x = 9y^2 \rightarrow x = 9(0)^2 = 0$

The values in Column A and B are equal.

Now, let's choose $y = 1 \rightarrow x = 9y^2 \rightarrow x = 9(1)^2 = 9$

Column A is greater. So, the relationship cannot be determined from the information given.

32) Answer: A.

Column A: $5x^2 - 4x + 14 = 5(1)^2 - 4(1) + 14 = 5 - 4 + 14 = 15$

Column B: $3x^3 + 2x^2 + 8 = 3(1)^3 + 2(1)^2 + 8 = 3 + 2 + 8 = 13$

33) Answer: D.

$4 > y > -2$

Let's choose some values for y. $y = -1$

Column A: $\frac{y}{5} = \frac{-1}{5}$

Column B: $\frac{5}{y} = \frac{5}{-1} = -5$

In this case, column A is bigger. $y = 1$

Column A: $\frac{y}{5} = \frac{1}{5}$

Column B: $\frac{5}{y} = \frac{5}{1} = 5$

In this case, Column B is bigger. So, the relationship cannot be determined from the information given.

34) Answer: D.

$\frac{a}{c} = \frac{d}{b}$ Here there are two equal fractions.

Let's choose some values for these variables. $\frac{1}{3} = \frac{3}{9}$

In this case, Column A is 2 (3 − 1) and Column B is 6 (9 − 3). Since, we can change the positions of these variables (for example put 3 for a and 9 for c), here relationship cannot be determined from the information given.

35) Answer: A.

First, let's find the number of digits when the printer prints 106 pages.

If there are 2 digits in each page and the printer prints 106 pages, then, there will be 212 digits. $106 \times 2 = 212$

However, we know that pages 1–9 have only one digit each, so we must subtract 9 from this total: $212 - 9 = 203$. We also know that the number 100th to 106th have three digits not two. So, we must add $7 \times 1 = 7$ digit (we counted two digits before) to this total: $203 + 7 = 210$.

It is given that 213 digits were printed, and we know that 106 pages results in 210 digits total, so there must be 107 total pages in the magazine. Column A is greater.

36) Answer: B.

The computer priced $126 includes 5% profit. Let x be the original cost of the computer. Then: $x + 5\% \ of \ x = 126 \rightarrow x + 0.05x = 126 \rightarrow 1.05x = 126 \rightarrow x = \frac{126}{1.05} = 120$

Column B is bigger.

37) Answer: A.

Column A: The largest number that can be written by rearranging the digits in 295 = 952

Column B: The largest number that can be written by rearranging the digits in 647=764

Answers and Explanations

ISEE - Upper Level

Practice Tests 1: Mathematics Achievement

1) Answer: A.

Simplify: $8 - 14 \div (7^2 \div 7) = 8 - 14 \div (49 \div 7) = 8 - 14 \div (7) = 8 - 2 = 6$

2) Answer: B.

$0.0002796 = \frac{2.796}{10,000} \Rightarrow 2.796 \times 10^{-4}$

3) Answer: C.

Write the proportion and solve for missing side.

$\frac{\text{Smaller triangle height}}{\text{Smaller triangle base}} = \frac{\text{Bigger triangle height}}{\text{Bigger triangle base}} \Rightarrow \frac{40cm}{120cm} = \frac{40+160cm}{x} \Rightarrow x = 600 \text{ cm}$

4) Answer: B.

$\frac{7}{31} = 0.225$ and $25\% = 0.25$ therefore x should be between 0.225 and 0.25

Only choice B ($\frac{15}{65} = 0.23$) is between 0.225 and 0.256

5) Answer: C.

Use FOIL (First, Out, In, Last) method.

$(x + 7)(x + 9) = x^2 + 7x + 9x + 63 = x^2 + 16x + 63$

6) Answer: D.

Solve for x.

$-3 \le 2x - 5 < 7 \Rightarrow$ (add 5 all sides) $-3 + 5 \le 2x - 5 + 5 < 7 + 5 \Rightarrow 2 \le 2x < 12$

$\Rightarrow$ (divide all sides by 2) $1 \le x < 6$

x is between 1 and 6. Choice D represent this inequality.

7) Answer: B.

First, divide all values on both sides of the equation by 4.

Then: $6x^2 + 21x + 9 = 0$

Use quadratic formula: $ax^2 + bx + c = 0$

$x_{1,2} = \frac{-b \pm \sqrt{b^2 - 4ac}}{2a}$

$6x^2 + 21x + 9 = 0 \quad \Rightarrow \quad$ then: $a = 6, b = 21$ and $c = 9$

$$x = \frac{-21 + \sqrt{21^2 - 4\times6\times9}}{2\times6} = -\frac{1}{2}$$

$$x = \frac{-21 - \sqrt{21^2 - 4\times6\times9}}{2\times6} = -3$$

8) Answer: D.

$\frac{1}{5}$ of the distance $8\frac{1}{8}$ miles is: $\frac{1}{5} \times 8\frac{1}{8} = \frac{1}{5} \times \frac{65}{8} = \frac{13}{8}$

Converting $\frac{13}{8}$ to a mixed number gives: $\frac{13}{8} = 1\frac{5}{8} = 1\frac{5}{8}$

9) Answer: C.

Simplify: $|8 - (45 \div |4 - 9|)| = |8 - (45 \div |-5|)| = |8 - (45 \div 5)| = |8 - 9| = |-1| = 1$

10) Answer: A.

The ratio of boy to girls is 3:5. Therefore, there are 3 boys out of 5 students. To find the answer, first divide the total number of students by 8, then multiply the result by 3.

$320 \div 8 = 40 \Rightarrow 40 \times 3 = 120$

11) Answer: D.

All Integers must end in one of the following digits:

0 when multiplied by itself ends in 0

1 when multiplied by itself ends in 1

2 when multiplied by itself ends in 4

3 when multiplied by itself ends in 9

4 when multiplied by itself ends in 6

5 when multiplied by itself ends in 5

6 when multiplied by itself ends in 6

7 when multiplied by itself ends in 9

8 when multiplied by itself ends in 4

9 when multiplied by itself ends in 1

Number 3,7,8 is not in the results.

12) Answer: B.

Translated 4 units down and 3 units to the left means:

$$(x, y) \Rightarrow (x - 3, y - 4)$$

13) Answer: B.

A linear equation is a relationship between two variables, x and y, and can be written in the form of $y = mx + b$

A non-proportional linear relationship takes on the form $y = mx + b$, where $b \neq 0$ and its graph is a line that does not cross through the origin.

Only in graph B, the line does not pass through the origin.

14) Answer: B.

Use Pythagorean theorem:

$$a^2 + b^2 = c^2 \to s^2 + h^2 = (7s)^2 \to s^2 + h^2 = 49s^2$$

Subtracting s^2 from both sides gives: $h^2 = 48s^2$

Square roots of both sides: $h = \sqrt{48s^2} = \sqrt{16 \times 3 \times s^2} = 4s\sqrt{3}$

15) Answer: D.

Let x be the number of purple marbles. Let's review the choices provided:

A. $\frac{1}{9}$, if the probability of choosing a purple marble is one out of ten, then:

$$Probability = \frac{number\ of\ desired\ outcomes}{number\ of\ total\ outcomes} = \frac{x}{15 + 20 + 45 + x} = \frac{1}{9}$$

Use cross multiplication and solve for x. $9x = 80 + x \to 8x = 80 \to x = 10$

Since, number of purple marbles can be 9, then, choice be the probability of randomly selecting a purple marble from the bag.

Use same method for other choices.

B. $\frac{3}{4}$; $\frac{x}{15 + 20 + 45 + x} = \frac{3}{4} \to 4x = 240 + 3x \to x = 240 \to x = 240$

C. $\frac{4}{9}$; $\frac{x}{15 + 20 + 45 + x} = \frac{4}{9} \to 9x = 320 + 4x \to 5x = 320 \to x = 64$

D. $\frac{5}{11}$; $\frac{x}{15 + 20 + 45 + x} = \frac{5}{11} \to 11x = 400 + 5x \to 6x = 400 \to x = 66.66$ (Number

of purple marbles cannot be a decimal).

16) Answer: C.

The area of the trapezoid is:

$$Area = \frac{1}{2}h(b_1 + b_2) = \frac{1}{2}(x)(12 + 18) = 120$$

$$\rightarrow 15x = 120 \rightarrow x = 8$$

$$y = \sqrt{6^2 + 8^2} = \sqrt{36 + 64} = \sqrt{100} = 10$$

The perimeter of the trapezoid is: $18 + 8 + 12 + 10 = 48$

17) Answer: A.

$$average = \frac{sum}{total} = \frac{54+55+47}{3} = \frac{156}{3} = 52$$

18) Answer: B.

Area of a rectangle = width × height

Area = $124 \times 53 = 6,572$

19) Answer: D.

$$76 \div \frac{1}{3} = 76 \times 3 = 228$$

20) Answer: C.

Let x be the original price of the dress. Then: 35% of $x = 26.25$

$$\frac{35}{100}x = 26.25$$

$$x = \frac{100 \times 26.25}{35} = 75$$

21) Answer: C.

$$\frac{9}{51} = 0.176$$

22) Answer: C.

60% of A is 1,200 Then: $0.6A = 1,200 \rightarrow A = \frac{1,200}{0.6} = 2,000$

15% of 2,000 is: $0.15 \times 2,000 = 300$

23) Answer: A.

$$(6.4 + 5.4 + 4.2) \times x = x$$

$$16x = x; \text{ Then } x = 0$$

24) Answer: B.

For sum of 4: (1 & 3) and (3 & 1) and (2 & 2), therefore we have 3 options.

For sum of 9: (3 & 6) and (6 & 3), (4 & 5) and (5 & 4), we have 4 options.

To get a sum of 5 or 10 for two dice: 3 + 4 = 7

Since, we have 6 × 6 = 36 total number of options, the probability of getting a sum of

4 or 9 is 7 out of 36 or $\frac{7}{36}$

25) Answer: D.

Simplify:

$$\frac{\frac{1}{3} - \frac{x+8}{6}}{\frac{x^2}{3} - \frac{5}{3}} = \frac{\frac{2}{6} - \frac{x+8}{6}}{\frac{x^2-5}{3}} = \frac{\frac{2-x-8}{6}}{\frac{x^2-5}{3}} = \frac{-x-6}{6} \times \frac{3}{x^2-5}$$

Then: $\frac{3(-x-6)}{6(x^2-5)} = \frac{-x-6}{2(x^2-5)} = \frac{-x-6}{2x^2-10}$

26) Answer: A.

The distance of A to B on the coordinate plane is: $\sqrt{(x_1 - x_2)^2 + (y_1 - y_2)^2} =$

$\sqrt{(11-6)^2 + (13-1)^2} = \sqrt{5^2 + 12^2} = \sqrt{25 + 144} = \sqrt{169} = 13$

The diameter of the circle is 13 and the radius of the circle is 6.5. Then: the

circumference of the circle is: $2\pi r = 2\pi(6.5) = 13\pi$

27) Answer: C.

$4x^2 - 18 = 46 \rightarrow 4x^2 = 64$

$x^2 = 16 \rightarrow x = \pm 4$

28) Answer: B.

Diameter = 8, then: Radius = 4

Area of a circle = $\pi r^2 \Rightarrow A = 3.14(4)^2 = 50.24$

29) Answer: D.

Write a proportion and solve. $\frac{8}{64} = \frac{x}{160} \rightarrow x = \frac{8 \times 160}{64} = 20$

30) Answer: C.

The mean of the data is 13. Then:

$\frac{x+7+18+17+12}{5} = 13 \rightarrow x + 54 = 65 \rightarrow x = 65 - 54 = 11$

31) Answer: B.

$82.36 \div 0.09 = 915.11$

32) Answer: D.

The difference of the file added, and the file deleted is:

$621,258 - 596,657 = 24,601$

$658,201 + 24,601 = 682,802$

33) Answer: B.

8 days 3 hours 20 minutes – 6 days 21 hours 35 minutes = 1 days and 5 hours and 45 min

34) Answer: A.

Formula of triangle area $= \frac{1}{2}$ (base × height)

Since the angles are 45-45-90, then this is an isosceles triangle, meaning that the base and height of the triangle are equal.

Triangle area $= \frac{1}{2}$ (base × height) $= \frac{1}{2}(14 \times 14) = 98$

35) Answer: D.

The area of the circle is 25π cm^2, then, its diameter is 10cm.

$area\ of\ a\ circle = \pi r^2 = 25\pi \rightarrow r^2 = 25 \rightarrow r = 5$

Radius of the circle is 5 and diameter is twice of it, 10.

One side of the square equals to the diameter of the circle. Then:

$Area\ of\ square = side \times side = 10 \times 10 = 100$

36) Answer: C.

80% of 15 = 12; 15 + 12 = 27

37) Answer: D.

$x^{\frac{1}{3}}$ equals to the root of x. Then: $28 + x^{\frac{1}{3}} = 31 \rightarrow 28 + \sqrt[3]{x} = 31 \rightarrow \sqrt[3]{x} = 3 \rightarrow x = 27$

$x = 27$ and $12 \times x$ equals: $12 \times 27 = 324$

38) Answer: D.

When a point is reflected over x axes, the (y) coordinate of that point changes to $(-y)$ while its x coordinate remains the same.

C $(4, 7) \rightarrow$ C' $(4, -7)$

39)Answer: D.

A set of ordered pairs represents y as a function of x if:

$x_1 = x_2 \rightarrow y_1 = y_2$

In choice A: $(4, -2)$ and $(4, 5)$ are ordered pairs with same x and different y, therefore y isn't a function of x.

In choice B: $(8, -2)$ and $(8, -3)$ are ordered pairs with same x and different y, therefore y isn't a function of x.

In choice C: $(-3, 6)$ and $(-3, 7)$ are ordered pairs with same x and different y, therefore y isn't a function of x.

40)Answer: C.

Only choice C represents the statement "triple the difference between 6 times H and 4 gives 48". $3(6H - 4) = 48$

41)Answer: C.

David's weekly salary is $400 plus 8% of $2,400. Then: $8\% \, of \, 2,400 =$

$0.08 \times 2,400 = 192$

$400 + 192 = 592$

42)Answer: D.

$9x^3y^4 + 5x^4y^5 - (7x^3y^4 - 5x^4y^5) =$

$9x^3y^4 + 5x^4y^5 - 7x^3y^4 + 5x^4y^5 = 2x^3y^4 + 10\,x^4y^5$

43)Answer: C.

Let P be circumference of circle A, then; $2\pi r_A = 24\pi \rightarrow r_A = 12$

$r_A = 4r_B \rightarrow r_B = \frac{12}{4} = 3 \rightarrow$ Area of circle B is; $\pi r_B^2 = 9\pi$

44)Answer: B.

Write a proportion and solve. $\frac{4}{5} = \frac{x}{60}$

Use cross multiplication: $5x = 240 \rightarrow x = 48$

45)Answer: C.

If the length of the box is 60, then the width of the box is one fifth of it, 12, and the height of the box is 2 (one sixth of the width). The volume of the box is:

$V = length \times width \times height = (60)(12)(2) = 1,440$

46) Answer: A.

The area of the square is 81 square inches. *Area of square* $= side \times side =$

$9 \times 9 = 81$

The length of the square is increased by 7 inches and its width decreased by 5 inches. Then, its area equals:

Area of rectangle $= width \times length = (9+7) \times (9-5) = 64$

The area of the square will be decreased by 17 square inches. $81 - 64 = 17$

47) Answer: B.

Number of squares equal to: $\frac{44 \times 12}{4 \times 4} = 11 \times 3 = 33$

Answers and Explanations

ISEE - Upper Level

Practice Tests 2: Quantitative Reasoning

1) Answer: B.

Area of the circle is less than 64 π. Use the formula of areas of circles.

Area $= \pi r^2 \Rightarrow 100\,\pi > \pi r^2 \Rightarrow 100 > r^2 \Rightarrow r < 10$

Radius of the circle is less than 10. Let's put 10 for the radius. Now, use the

circumference formula: Circumference $= 2\pi r = 2\pi\,(10) = 20\,\pi$

Since the radius of the circle is less than 10. Then, the circumference of the circle

must be less than 20 π. Only choice B is less than 20π

2) Answer: C.

Let x be all expenses, then $\frac{18}{100}x = \$360 \rightarrow x = \frac{100 \times \$360}{18} = 2{,}000$

He spent for his rent: $\frac{32}{100} \times \$2{,}000 = \$640$

3) Answer: D.

There is not enough information to determine the answer of the question. An average

speed represents a distance divided by time and it does not provide information about

the speed at specific time. Etta could drove exactly 64 miles per hour from start to

finish, or she could drive 60 miles per hour for half of distance and 62 miles per hour

for the other half.

4) Answer: B.

$(4x + 9) - (4x - 5) = 4x - 4x + 9 + 5 = 14$

5) Answer: C.

The percentage discount is the reduction in price divided by the original price. The

difference between original price and sale price is: $\$18.32 - \$8.98 = \$9.34$

The percentage discount is this difference divided by the original price:

$\$9.34 \div \$18.32 \cong 0.5098$

Convert the decimal to a percentage by multiplying by 100%:

$0.5098 \times 100\% = 50.98\%$

6) Answer: C.

The smallest possible value of $f(x)$ will occur when $x = 0$. Since x^2 is always positive, any positive or negative value of x will make the value of $f(x)$ greater than 12. Substitute 0 for x and evaluate the expression: $f(0) = 2(0)^2 + 12 = 12$

7) Answer: D.

There are 2 sets of values, one set from 1 to $n + 1$, and the other set from $n + 2$ to $2n$. Since the second set begins immediately after the first set, the two sets can be combined. The sum of the positive integers from 1 to $2n$ inclusive is equal to the sum of the positive integers from 1 to $n + 1$ plus the sum of the positive integers from $n + 2$ to $2n$: $3,875 + 9,328 = 13,203$

8) Answer: C.

If two triangles are similar, then the ratios of corresponding sides are equal.

$\frac{AC}{AE} = \frac{BC}{DE} = \frac{36}{9} = 4 \implies \frac{AC}{AE} = 4$

This ratio can be used to find the length of AC:

$AC = 4 \times AE \implies AC = 4 \times 4 \implies AC = 16$

The length of AE is given as 4 and we now know the length of AC is 16, therefore:

$EC = AC - AE \implies EC = 16 - 4 \implies EC = 12$

9) Answer: C.

Let's review the choices provided:

A. number of books sold in June is: 180

   One third the number of books sold in July is: $\frac{750}{3} = 250 \rightarrow 180 < 250$

B. number of books sold in May is: 1,200

   One third the number of books sold in May is: $\frac{1,200}{3} = 400 \rightarrow 400 \neq 750$

C. Number of books sold in April is: 540

   Number of books sold in June is: 180 $\rightarrow \frac{540}{180} = 3$

D. $540 + 180 = 720 < 750$

10) Answer: D.

Find the value of each choice:

A. $2 \times 3 \times 3 \times 3 = 54$

B. $2 \times 2 \times 5 \times 7 = 140$

C. $3 \times 7 = 21$

D. $2 \times 2 \times 3 \times 3 \times 5 = 180$

11) Answer: A.

The median of a set of data is the value located in the middle of the data set. Combine the 2 sets provided, and organize them in ascending order:

$\{1, 3, 4, 7, 8, 10, 12, 15, 19, 21\}$

Since there are an even number of items in the resulting list, the median is the average of the two middle numbers.

Median $= (8 + 10) \div 2 = 9$

12) Answer: D.

Write a proportion and solve.

$\frac{12in}{80feet} = \frac{36in}{x} \rightarrow x = \frac{80 \times 36}{12} = 240 \; feet$

13) Answer: D.

Let x be the number of current stamps in the collection. Then:

$\frac{8}{7}x - x = 60 \rightarrow \frac{1}{7}x = 60 \rightarrow x = 420$

20% more of 420 is: $420 + 0.20 \times 420 = 420 + 84 = 504$

14) Answer: B.

Let x be number of gallons the tank can hold when it is full. Then:

$\frac{3}{7}x = 21 \rightarrow x = \frac{3}{7} \times 21 = 9$

15) Answer: B.

Recall that the formula for the average is:

$\text{Average} = \frac{sum \; of \; data}{number \; of \; data}$

First, compute the total weight of all balls in the basket:

$$15 \text{ g} = \frac{total\ weight}{24\ balls}$$

$15 \times 24 = $ total weight $= 360$ g

Next, find the total weight of the 6 heaviest balls:

$$18 \text{ g} = \frac{total\ weight}{6\ marbles}$$

$18 \text{ g} \times 6 = $ total weight $= 108$ g

The total weight of six heaviest balls is 108 g. Then, the total weight of the remaining 15 balls is 252 g. 360 g – 108 g = 252 g.

The average weight of the remaining balls:

$$\text{Average} = \frac{252\ g}{18\ marbles} = 14 \text{ g per ball}$$

16) Answer: D.

If 12 balls are removed from the bag at random, there will be two ball in the bag. The probability of choosing a brown ball is 2 out of 18. Therefore, the probability of not choosing a brown ball is 12 out of 18 and the probability of having not a brown ball after removing 12 balls is the same.

17) Answer: A.

$(4x^2 - y^2) = (2x - y)(2x + y)$

Then: $(4x^2 - y^2) = 8 \times 4 = 32$

18) Answer: D.

The ratio of boy to girls is 4: 7. Therefore, there are 4 boys out of 11 students. To find the answer, first divide the total number of students by 11, then multiply the result by 4.

$77 \div 11 = 7 \Rightarrow 7 \times 4 = 28$

There are 28 boys and 49 (77 – 28) girls. So, 21 more boys should be enrolled to make the ratio 1:1

19) Answer: A.

The formula for the area of a rectangle is: Area = Width × Length

It is given that L = 4W and that A = 256. Substitute the given values into our equation and solve for W:

$256 = w \times 4w$

$256 = 4w^2 \Rightarrow w^2 = 64 \Rightarrow w = 8$

It is given that $L = 4W$, therefore, $L = 4 \times 8 = 32$

The perimeter of a rectangle is: $2L + 2W$

Perimeter $= 2 \times 32 + 2 \times 8$,   Perimeter $= 80$

20) Answer: D.

The sum of 7 numbers is greater than 350 and less than 420. Then, the average of the 7 numbers must be greater than 50 and less than 60.

$$\frac{350}{7} < x < \frac{420}{7} \Rightarrow 50 < x < 60$$

The only choice that is between 50 and 60 is 55.

21) Answer: B.

In order to solve for the variable b, first take square roots on both sides:

$\sqrt{a} = \frac{2cz}{3b}$, then multiply both sides by b: $3b\sqrt{a} = 2cz$

Now, divide both sides by c: $Z = \frac{3b\sqrt{a}}{2c}$

22) Answer: A.

The average is the sum of all terms divided by the number of terms.

$14 + 28 + 18 = 60$

$60 \div 3 = 20$

This is greater than 18.

23) Answer: B.

Since both columns have 268 as a factor, we can ignore that number.

$14 \times 15 = 210$; $17 \times 15 = 255$; Column B is greater

24) Answer: D.

Since x is an integer and can be positive and negative, then the relationship cannot be determined from the information given. Let's choose some values for x.

$x = 1$, then the value in column A is greater. $1 > \frac{1}{-7}$

Let's choose a negative value for x.

$x = -1$, then the value in column B is greater. $-1 < \frac{-1}{-7} \rightarrow -1 < \frac{1}{7}$

25) Answer: A.

First, find the values of x in both columns.

Column A: $4|3x - 8| = 4 \rightarrow |3x - 8| = 1$

$3x - 8$ can be 1 or -1.

$3x - 8 = 1 \rightarrow 3x = 9 \rightarrow x = 3$

$3x - 8 = -1 \rightarrow 3x = 7 \rightarrow x = \frac{7}{3}$

Column B: $4|3x + 8| = 4 \rightarrow |2x + 6| = 1$

$3x + 8$ can be 1 or -1.

$3x + 8 = 1 \rightarrow 3x = -7 \rightarrow x = -\frac{7}{3}$

$3x + 8 = -1 \rightarrow 3x = -10 \rightarrow x = -\frac{10}{3}$

The greatest value of x in column A is 3 and the greatest value of x in columns B is $-\frac{7}{3}$.

26) Answer: D.

Simplify both columns.

Column A: $(x)^4(x)^2 = x^6$

Column B: $(x^4)^2 = x^8$

Column A evaluates to x^6 and Column B evaluates to x^8. In the case where $x = 0$, the two columns will be equal, but if $x = 2$, the two columns will not be equal. Consequently, the relationship cannot be determined.

27) Answer: C.

Column A: $2^3 = 8$

Column B: $\sqrt[3]{521} = 8$ (recall that $8^3 = 512$)

28) Answer: A.

A number raised to the exponent $(\frac{1}{2})$ is the same thing as evaluating the square root of the number. Therefore: $(24)^{\frac{1}{2}} = \sqrt{24}$

Since $\sqrt{36}$ is greater than $\sqrt{24}$, column A ($\sqrt{36} = 6$) is greater than $\sqrt{42}$.

29) Answer: A.

Let x be the original price of the sport jacket. The selling price of a sport jacket including 40% discount is \$180. Then:

$$x - 0.40x = 180 \rightarrow 0.60x = 180 \rightarrow x = \frac{180}{0.60} = 300$$

The original price of the jacket is \$300 which is greater than column B (\$275).

30) Answer: B.

The value of x has to be less than 22, which is less than 22. Column B is greater.

Recall that when the positive powers of numbers between 0 and 1 increases, the value of the number decrease. For example: $(0.6)^2 > (0.6)^3 \rightarrow 0.36 > 0.216$

So, $(0.55)^{33} > (0.55)^{34}$

31) Answer: D.

The probability that an event will occur + the probability that that event will NOT occur must equal 1. Since we don't have any numerical information about the probability, it is possible that the probability that event x occurs is 25%, 50% or any other percent. The probability that event x will not occur will always be 100% minus the probability that event x does occur. Because both columns can exhibit a range of values, the relationship cannot be determined.

32) Answer: B.

Because of the word "and" the events described in each column must be calculated separately and then multiplied:

For column A: Probability of rolling a 2: $\frac{1}{6}$

Probability of getting heads: $\frac{1}{2}$

$\frac{1}{6} \times \frac{1}{2} = \frac{1}{12}$

For column B: Probability of an odd number: $\frac{3}{6} = \frac{1}{2}$

Probability of getting a spades: $\frac{13}{52} = \frac{1}{4}$

$\frac{1}{2} \times \frac{1}{4} = \frac{1}{8}$

Since $\frac{1}{8}$ is a larger number than $\frac{1}{12}$, Colum B is greater

33) Answer: C.

To raise a quantity to a negative power, invert the numerator and denominator, and then raise the base to the indicated power. Therefore:

$(\frac{5}{1})^{-3} = (\frac{1}{5})^3$; The Columns are the same value.

34) Answer: D.

First, simplify the inequality:

$$3x + 3 > 6x \rightarrow 3 > 3x \rightarrow \frac{3}{3} > x \rightarrow 1 > x$$

Since x is less than 1, and x can be 0 (greater than -1) or -2 (less than -1), the relationship cannot be determined.

35) Answer: A.

Sum of five quarter, three nickels, and four pennies is: 5($0.25) + 3($0.05) + 4($0.01) = $1.44

36) Answer: A.

Let's consider the properties of odd and even integers:

Odd +/− Odd = Even   Odd × Odd = Odd

Even+/−Even= Even   Even × Even = Even

Odd +/− Even = Odd   Odd × Even = Even

For <u>Column A</u>: $(y)(x - y)$

(odd) (even +odd) $\Rightarrow$ (odd) (odd) $\Rightarrow$ even

Now let's review the columns. For <u>column B</u>: $(x - y)^2 + y$

(even −odd) ^{2} + odd

(even)^2 − odd $\Rightarrow$ (even)(even) − odd $\Rightarrow$ even− odd $\Rightarrow$ odd

Since an even number is considered greater according to the problem statement, the answer is A.

37) Answer: D.

Factor the expression if possible. Begin by moving all terms to one side before factoring:

$x^2 - 2x - 16 = 32 \Rightarrow x^2 - 2x - 48 = 0$

To factor this quadratic, find two numbers that multiply to -48 and sum to -2:

$(x - 8)(x + 6) = 0$

Set each expression in parentheses equal to 0 and solve:

$x - 8 = 0 \Rightarrow x = 8$

$x + 6 = 0 \Rightarrow x = -6$

Quadratic equations can have TWO possible solutions. Since one of these is greater than 0 and one of them is less than 0, we cannot determine the relationship between the columns.

Answers and Explanations

ISEE - Upper Level

Practice Tests 2: Mathematics Achievement

1) Answer: B.

$$\frac{13}{26} = 0.5$$

2) Answer: C.

Let x be the integer. Then: $3x - 10 = 29$

Add 10 both sides: $3x = 39 \Rightarrow$ Divide both sides by 3: $x = 13$

3) Answer: D.

Plug in each pair of numbers in the equation. The answer should be 11.

   A. $(3, -1)$: $3(3) + 4(-1) = 5$ No!

   B. $(-3, 1)$: $3(-3) + 4(1) = -5$ No!

   C. $(-2, 0)$: $3(-2) + 4(0) = -6$ No!

   D. $(2, 1)$: $3(2) + 5(1) = 11$ Yes!

4) Answer: C.

The sum of supplement angles is 180. Let x be that angle. Therefore, $x + 8x = 180$

$9x = 180$, divide both sides by 9: $x = 20$

5) Answer: B.

$x\% \ 40 = 3.2$

$$\frac{x}{100} \ 40 = 3.2 \rightarrow x = \frac{3.2 \times 100}{40} = 8$$

6) Answer: D.

Use Pythagorean Theorem: $a^2 + b^2 = c^2$

$10^2 + 24^2 = c^2 \Rightarrow 676 = c^2 \Rightarrow c = 26$

7) Answer: B.

Subtract $\frac{1}{4b}$ and $\frac{1}{b^2}$ from both sides of the equation. Then:

$$\frac{1}{4b^2} + \frac{1}{4b} = \frac{1}{2b^2} \rightarrow \frac{1}{4b^2} - \frac{1}{2b^2} = -\frac{1}{4b}$$

Multiply both numerator and denominator of the fraction $\frac{1}{b^2}$ by 2.

Then: $\frac{1}{4b^2} - \frac{2}{4b^2} = -\frac{1}{4b}$

Simplify the first side of the equation: $-\frac{2}{4b^2} = -\frac{1}{2b}$

Use cross multiplication method: $4b = 4b^2 \rightarrow 4 = 4b \rightarrow b = 1$

8) Answer: B.

First, multiply both sides of inequality by 5. Then:

$\frac{|7+x|}{3} \leq 6 \rightarrow |7 + x| \leq 18 \rightarrow -18 \leq 7 + x \leq 18$

Since $7 + x$ can be positive or negative, then:

$7 + x \leq 18 \; or \; 7 + x \geq -18$

Then: $x \leq 11 \; or \; x \geq -25$

Choice B is correct.

9) Answer: B.

Plug in the value of $x = 8$ into both equations. Then:

$C(x) = x^2 + 2x = (8)^2 + 2(8) = 64 + 16 = 80$

$R(x) = 15x = 15 \times 8 = 120$

$120 - 80 = 40$

So, the profit is \$40.

10) Answer: A.

From choices provided, only choice D is correct.

$E = A + 5$

$A = S - 4 \Rightarrow S = A + 4$

11) Answer: D.

Simplify. $9x^3y^2(2x^2y)^3 = 9x^3y^2(8x^6y^3) = 72x^9y^5$

12) Answer: B.

28 hours = 100,800 seconds

3,330 minutes = 252,600 seconds

1 days and 2 hours = 26 hours = 93,600 seconds

13) Answer: C.

Let's review the choices provided:

A. $(4 \times 10^3) + (6 \times 10^2) + (7 \times 10) = 4,000 + 600 + 70 = 4,670$

B. $(4 \times 10^2) + (6 \times 10^1) - 10 = 400 + 60 - 10 = 450$

C. $(4 \times 10^2) + (6 \times 10^1) + 7 = 400 + 60 + 7 = 467$

D. $(4 \times 10^1) + (6 \times 10^2) + 75 = 40 + 600 + 7 = 647$

Only choice C equals to 467.

14) Answer: C.

$\frac{4}{400} = \frac{x}{540} \Rightarrow x = \frac{4 \times 540}{400} = 5.4$

15) Answer: B.

Area of a circle equals: $A = \pi r^2$

The new diameter is 70% larger than the original then the new radius is 30% larger than the original (a diameter is twice of a radius).

30% larger than r is $1.3r$. Then, the area of larger circle is:

$A = \pi r^2 = \pi (1.3r)^2 = \pi (1.69r^2) = 1.69\pi r^2$

$1.69\pi r^2$ is 69% bigger than πr^2.

16) Answer: B.

$84.24 \div 0.08 = 1,053$

17) Answer: D.

$4\frac{2}{3} - 3\frac{5}{6} = 4\frac{4}{6} - 3\frac{5}{6} = \frac{28}{6} - \frac{23}{6} = \frac{5}{6}$

18) Answer: C.

Let's review the choices provided. Put the values of x and y in the equation.

A. $(0, -5)$ $\Rightarrow x = 0 \Rightarrow y = -5$ This is true!

B. $(-1, -12)$ $\Rightarrow x = -1 \Rightarrow y = -12$ This is true!

C. $(-3, 8)$ $\Rightarrow x = -3 \Rightarrow y = -26 \neq 8$ This is not true!

D. $(1, 2)$ $\Rightarrow x = 1 \Rightarrow y = 2$ This is true!

19) Answer: B.

$C = 2\pi r = \pi d \rightarrow C = \pi \times 10 = 10\pi$

$\pi = 3.14 \rightarrow C = 10\pi = 31.4 = 31$

20) Answer: B.

The area of the non-shaded region is equal to the area of the bigger rectangle

subtracted by the area of smaller rectangle.

Area of the bigger rectangle $= 14 \times 16 = 224$

Area of the smaller rectangle $= 11 \times 8 = 88$

Area of the non-shaded region $= 224 - 88 = 136$

21) Answer: A.

First draw an isosceles triangle. Remember that two

sides of the triangle are equal.

Let put a for the legs. Then: Use Pythagorean theorem

to find the value of a:

$a^2 + b^2 = c^2 \rightarrow a^2 + a^2 = 10^2$

Isosceles right triangle

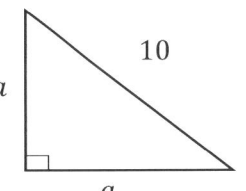

Simplify: $2a^2 = 100 \rightarrow a^2 = 50 \rightarrow a = \sqrt{50}$

$a = \sqrt{50} \Rightarrow$ area of the triangle is $= \frac{1}{2}\left(\sqrt{50} \times \sqrt{50}\right) = \frac{1}{2} \times 50 = 25 \; cm^2$

22) Answer: C.

To find total number of miles driven by Ed that week, you only need to subtract

$42,857$ from $43,128$. $43,128 - 42,857 = 271$

23) Answer: D.

To find the maximum value of y, the expression $(x - 3)^2$ must be equal to 0.

Because it has a negative sign. Since $x - 3$ is to the power of 3, it cannot be negative.

To get 0 for the expression $(x - 3)^2$, x must be 3.

Plug in 3 for x in the equation: $y = -(x - 3)^2 + 8 \rightarrow y = -(3 - 3)^2 + 8 = 8 \Rightarrow$

The maximum value of y is 8.

24) Answer: B.

$\begin{cases} -3x + y = -5 \\ 4x + 3y = 11 \end{cases} \Rightarrow$ Multiplication (-3) in first equation $\Rightarrow \begin{cases} 9x - 3y = 15 \\ 4x + 3y = 11 \end{cases}$

Add two equations together $\Rightarrow 13x = 26 \Rightarrow x = 2$ then: $y = 1$

25) Answer: A.

First draw an isosceles triangle. Remember that two sides of the triangle are equal.

Let put a for the legs. Then:

$a = 4 \Rightarrow$ area of the triangle is $=$

$\frac{1}{2}(4 \times 4) = \frac{16}{2} = 8 \ cm^2$

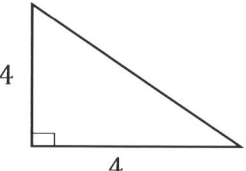

Isosceles right triangle

26) Answer: A.

Factor each trinomial $x^2 + 4x - 45$ and $x^2 - 12x + 35$

$x^2 + 4x - 45 \Rightarrow (x-5)(x+9)$

$x^2 - 12x + 35 \Rightarrow (x-7)(x-5)$

The common factor of both expressions is $(x-5)$.

27) Answer: C.

$x + y = 18$

Then: $2x + 2y = 2(x+y) = 2 \times 18 = 36$

28) Answer: C.

$\frac{350}{14} = 25$

29) Answer: C.

Use FOIL (First, Out, In, Last)

$(5x + 5)(x + 4) = 5x^2 + 20x + 5x + 20 = 5x^2 + 25x + 20$

30) Answer: B.

Plug in the values of x and y in the equation:

$3 \blacksquare 4 = \sqrt{3^2 + 4(4)} = \sqrt{9 + 16} = \sqrt{25} = 5$

31) Answer: D.

Let x be the capacity of one tank. Then, $\frac{4}{7}x = 210 \rightarrow x = \frac{210 \times 4}{7} = 120$ Liters

The amount of water in three tanks is equal to: $2 \times 120 = 240$ Liters

32) Answer: B.

$$\text{Average} = \frac{\text{sum of terms}}{\text{number of terms}}$$

The sum of the weight of all girls is: $16 \times 45 = 720\ kg$

The sum of the weight of all boys is: $24 \times 55 = 1,320\ kg$

The sum of the weight of all students is: $720 + 1,320 = 2,040\ kg$

$$\text{Average} = \frac{2,040}{40} = 51$$

33) Answer: B.

$$\frac{35}{100} \times 720 = x \Rightarrow x = 252$$

34) Answer: B.

$$\begin{array}{r} 28\ \text{hr.}\quad 15\ \text{min.} \\ -\ 19\ \text{hr.}\quad 37\ \text{min.} \\ \hline 8\ \text{hr}\ and\ 38\text{min} \end{array}$$

35) Answer: A.

First, convert all measurement to foot.

One foot is 12 inches. Then: $4\ \text{inches} = \frac{4}{12} = \frac{1}{3}\ \text{feet}$

The volume flower box is: $\text{length} \times \text{width} \times \text{height} = 5 \times \frac{1}{3} \times 6 = 10$ cubic feet.

36) Answer: B.

Let's write equations based on the information provided:

Michelle = Karen − 9 ⇒ Karen= Michelle + 9

Michelle = David − 3 ⇒ David = Michelle + 3

Karen + Michelle + David = 84

Now, replace the ages of Karen and David by Michelle. Then:

Michelle + 9 + Michelle + Michelle + 3 = 84

3Michelle + 12 = 84 ⇒ 3 Michelle = 84 − 12

3 Michelle = 72 ⇒ Michelle = 24

37) Answer: D.

Use interest rate formula:

$$Interest = principal \times rate \times time = 950 \times 0.02 \times 2 = 38$$

38) Answer: D.

Mason travels $\frac{2}{7}$ of 42 hours. $\frac{2}{7} \times 42 = 12$

Mason will be on the road for 12 hours.

39) Answer: D.

To find the discount, multiply the number by (100% – rate of discount).

Therefore, for the first discount we get: $(570)(100\% - 30\%) = (570)(0.70)$

For the next 18 % discount: $(570)(0.70)(0.82)$

40) Answer: D.

$\frac{x^4}{18} \Rightarrow$ reciprocal is: $\frac{18}{x^4}$

41) Answer: B.

Plug in the values of x and y in the expression:

$2x^2(y + 6) = 2(0.3)^2(4 + 6) = 2(0.09)(10) = 1.8$

42) Answer: C.

$A = bh \Rightarrow A = 4 \times 5.6 = 22.4$

43) Answer: B.

The ratio of boys to girls is 7:5. Therefore, there are 7 boys out of 12 students. To find the answer, first divide the number of boys by 7, then multiply the result by 12.

$280 \div 7 = 40 \Rightarrow 40 \times 12 = 480$

44) Answer: A.

Write a proportion and solve for the missing number.

$\frac{32}{14} = \frac{8}{x} \rightarrow 32x = 8 \times 14 = 112$

$32x = 112 \rightarrow x = \frac{112}{32} = 3.5$

45) Answer: B.

To find the area of the shaded region subtract smaller circle from bigger circle.

$S_{bigger} - S_{smaller} = \pi(r_{bigger})^2 - \pi(r_{smaller})^2 \Rightarrow S_{bigger} - S_{smaller} =$

$\pi(8)^2 - \pi(3)^2 \Rightarrow 64\pi - 9\pi = 55\pi$

46) Answer: C.

To add two matrices, first we need to find corresponding members from each matrix.

$$\begin{vmatrix} 3 & 5 \\ -2 & -2 \\ -8 & 6 \end{vmatrix} + \begin{vmatrix} 1 & -1 \\ 5 & -1 \\ 7 & -4 \end{vmatrix} = \begin{vmatrix} 4 & 4 \\ 3 & -3 \\ -1 & 2 \end{vmatrix}$$

47) Answer: C.

The area of 7 feet by 7 feet room is 49 square feet.

$7 \times 7 = 49$

"End"

Printed in Great Britain
by Amazon

67899941R00138